STUDENT'S SOLUTIONS MANUAL

CONTEMPORARY BUSINESS MATHEMATICS WITH CANADIAN APPLICATIONS

Sixth Edition

S.A. Hummelbrunner

Prentice
Hall

Toronto

0-13-089263-7

Acquisitions Editor: Dave Ward
Developmental Editor: Madhu Ranadive
Production Editor: Mary Ann McCutcheon
Production Coordinator: Janette Lush

5   05 04 03

Printed and bound in Canada

Contents

Part One: Mathematics Fundamentals   1

1   Review of arithmetic

Exercise 1.1   1
Exercise 1.2   1
Exercise 1.3   3
Exercise 1.4   7
Exercise 1.5   8
Exercise 1.6   10
Review Exercise   11
Self-Test   15

2   Review of basic algebra

Exercise 2.1   17
Exercise 2.2   18
Exercise 2.3   20
Exercise 2.4   21
Exercise 2.5   22
Exercise 2.6   23
Exercise 2.7   25
Review Exercise   26
Self-Test   33

3   Ratio, proportion and percent

Exercise 3.1   36
Exercise 3.2   38
Exercise 3.3   39
Exercise 3.4   43
Exercise 3.5   45
Exercise 3.6   47
Exercise 3.7   48
Exercise 3.8   49
Exercise 3.9   49
Review Exercise   49
Self-Test   54

4   Linear systems

      Exercise 4.1   57
      Exercise 4.2   60
      Exercise 4.3   64
      Exercise 4.4   67
      Review Exercise   69
      Self-Test   75

Part Two: Mathematics of Business and Management   78

5   Business applications – Depreciation and break-even analysis

      Exercise 5.1   78
      Exercise 5.2   86
      Review Exercise   88
      Self-Test   92

6   Commercial discount, markup and markdown

      Exercise 6.1   93
      Exercise 6.2   94
      Exercise 6.3   95
      Exercise 6.4   97
      Exercise 6.5   102
      Review Exercise   107
      Self-Test   111

7   Simple interest

      Exercise 7.1   113
      Exercise 7.2   113
      Exercise 7.3   113
      Exercise 7.4   114
      Exercise 7.5   116
      Exercise 7.6   117
      Exercise 7.7   118
      Review Exercise   121
      Self-Test   124

8   Simple interest applications

Exercise 8.1	126
Exercise 8.2	127
Exercise 8.3	127
Exercise 8.4	128
Exercise 8.5	128
Exercise 8.6	129
Exercise 8.7	131
Exercise 8.8	132
Exercise 8.9	133
Review Exercise	134
Self-Test	137

Part Three: Mathematics of Finance and Investment   140

9   Compound interest – Future value and present value

Exercise 9.1	140
Exercise 9.2	141
Exercise 9.3	145
Exercise 9.4	146
Exercise 9.5	148
Review Exercise	152
Self-Test	155

10   Compound interest – Further topics

Exercise 10.1	156
Exercise 10.2	160
Exercise 10.3	163
Exercise 10.4	167
Review Exercise	169
Self-Test	173

11   Ordinary simple annuities

Exercise 11.1	175
Exercise 11.2	175
Exercise 11.3	176
Exercise 11.4	178
Exercise 11.5	180
Review Exercise	182
Self-Test	184

12   Other simple annuities

Exercise 12.1   185
Exercise 12.2   186
Exercise 12.3   187
Exercise 12.4   189
Exercise 12.5   190
Review Exercise   193
Self-Test   196

13   Annuities – Finding R, n, or I

Exercise 13.1   198
Exercise 13.2   204
Exercise 13.3   211
Review Exercise   215
Self-Test   224

14   Amortization of debts

Exercise 14.1   227
Exercise 14.2   233
Exercise 14.3   240
Exercise 14.4   244
Review Exercise   248
Self-Test   257

15   Bond valuation and sinking funds

Exercise 15.1   260
Exercise 15.2   263
Exercise 15.3   267
Exercise 15.4   269
Exercise 15.5   270
Review Exercise   276
Self-Test   285

16   Investment decision applications

Exercise 16.1   289
Exercise 16.2   290
Exercise 16.3   293
Review Exercise   295
Self-Test   299

PART ONE *Mathematics fundamentals*

1 Review of arithmetic

Exercise 1.1

A. 1. $12 + 6 \div 3 = 12 + 2 = 14$

3. $(3 \times 8 - 6) \div 2 = (24 - 6) \div 2 = 18 \div 2 = 9$

5. $(7 + 4) \times 5 - 2 = 11 \times 5 - 2 = 55 - 2 = 53$

7. $5 \times 3 + 2 \times 4 = 15 + 8 = 23$

9. $(3 \times 9 - 3) \div 6 = (27 - 3) \div 6 = 24 \div 6 = 4$

11. $6(7 - 2) - 3(5 - 3) = 6(5) - 3(2) = 30 - 6 = 24$

13. $\dfrac{16-8}{8-2} = \dfrac{8}{6} = \dfrac{4}{3} = 1.3333333$

15. $4(8 - 5)^2 - 5(3 + 2^2) = 4(3)^2 - 5(3 + 4) = 4(9) - 5(7) = 36 - 35 = 1$

Exercise 1.2

A. 1. $\dfrac{24}{36} = \dfrac{24{:}2}{36{:}2} = \dfrac{12}{18} = \dfrac{12{:}2}{18{:}2} = \dfrac{6}{9} = \dfrac{6{:}3}{9{:}3} = \dfrac{2}{3};$ also $\dfrac{24{:}12}{36{:}12} = \dfrac{2}{3}$

3. $\dfrac{210}{360} = \dfrac{210{:}10}{360{:}10} = \dfrac{21}{36} = \dfrac{21{:}3}{36{:}3} = \dfrac{7}{12};$ also $\dfrac{210{:}30}{360{:}30} = \dfrac{7}{12}$

5. $\dfrac{360}{225} = \dfrac{360{:}5}{225{:}5} = \dfrac{72}{45} = \dfrac{72{:}9}{45{:}9} = \dfrac{8}{5};$ also $\dfrac{360{:}45}{225{:}45} = \dfrac{8}{5}$

7. $\dfrac{144}{360} = \dfrac{144{:}2}{360{:}2} = \dfrac{72}{180} = \dfrac{72{:}9}{180{:}9} = \dfrac{8}{20} = \dfrac{8{:}4}{20{:}4} = \dfrac{2}{5};$ also $\dfrac{144{:}72}{360{:}72} = \dfrac{2}{5}$

9. $\dfrac{25}{365} = \dfrac{25{:}5}{365{:}5} = \dfrac{5}{73}$

11. $\dfrac{365}{73} = \dfrac{365:73}{73:73} = \dfrac{5}{1}$

B. 1. $\dfrac{11}{8} = 1.375$

3. $\dfrac{5}{3} = 1.6666667 = 1.\dot{6}$

5. $\dfrac{11}{6} = 1.8333333 = 1.8\dot{3}$

7. $\dfrac{13}{12} = 1.0833333 = 1.08\dot{3}$

C. 1. $3\dfrac{3}{8} = 3.375$

3. $8\dfrac{1}{3} = 8.3333333$

5. $33\dfrac{1}{3} = 33.3333333$

7. $7\dfrac{7}{9} = 7.7777778$

D. 1. 5.63

3. 18.00

5. 57.70

7. 13.00

E. 1. $\dfrac{54}{0.12 \times \frac{225}{360}} = \dfrac{54}{0.12 \times 0.625} = \dfrac{54}{0.075} = 720$

3. $620\left(1 + 0.14 \times \dfrac{45}{360}\right) = 620(1+0.0175) = 620(1.0175) = 630.85$

5. $2100\left(1 - 0.135 \times \dfrac{240}{360}\right) = 2100(1 - 0.09) = 2100(0.91) = 1911$

7. $\dfrac{250\,250}{1 + 0.15 \times \frac{330}{360}} = \dfrac{250\,250}{1 + 0.1375} = \dfrac{250\,250}{1.1375} = 220\,000$

9. $\dfrac{3460}{1 - 0.18 \times \frac{270}{360}} = \dfrac{3460}{1 - 0.135} = \dfrac{3460}{0.865} = 4000$

Exercise 1.3

A. 1. $64\% = \dfrac{64}{100} = 0.64$

3. $2.5\% = \dfrac{2.5}{100} = 0.025$

5. $0.5\% = \dfrac{0.5}{100} = 0.005$

7. $250\% = \dfrac{250}{100} = 2.5$

9. $450\% = \dfrac{450}{100} = 4.5$

11. $0.9\% = \dfrac{0.9}{100} = 0.009$

13. $6.25\% = \dfrac{6.25}{100} = 0.0625$

15. $99\% = \dfrac{99}{100} = 0.99$

17. $0.05\% = \dfrac{0.05}{100} = 0.0005$

19. $\dfrac{1}{2}\% = \dfrac{0.5}{100} = 0.005$

21. $9\dfrac{3}{8}\% = \dfrac{9.375}{100} = 0.09375$

23. $162\dfrac{1}{2}\% = \dfrac{162.5}{100} = 1.625$

25. $\dfrac{1}{4}\% = \dfrac{0.25}{100} = 0.0025$

27. $1\dfrac{3}{4}\% = \dfrac{1.75}{100} = 0.0175$

29. $137\dfrac{1}{2}\% = \dfrac{137.5}{100} = 1.375$

31. $0.875\% = \dfrac{0.875}{100} = 0.00875$

33. $33\dfrac{1}{3}\% = \dfrac{33.\overline{3}}{100} = 0.\overline{3}$

35. $16\dfrac{2}{3}\% = \dfrac{16.\overline{6}}{100} = 0.1\overline{6}$

37. $183\dfrac{1}{3}\% = \dfrac{183.\overline{3}}{100} = 1.8\overline{3}$

39. $133\dfrac{1}{3}\% = \dfrac{133.\overline{3}}{100} = 1.\overline{3}$

B. 1. $25\% = \dfrac{25}{100} = \dfrac{1}{4}$

3. $175\% = \dfrac{175}{100} = \dfrac{7}{4}$

5. $37\dfrac{1}{2}\% = \dfrac{37.5}{100} = \dfrac{375}{1000} = \dfrac{3}{8}$

4

7. $4\% = \dfrac{4}{100} = \dfrac{1}{25}$

9. $8\% = \dfrac{8}{100} = \dfrac{2}{25}$

11. $40\% = \dfrac{40}{100} = \dfrac{2}{5}$

13. $250\% = \dfrac{250}{100} = \dfrac{5}{2}$

15. $12\dfrac{1}{2}\% = \dfrac{12.5}{100} = \dfrac{125}{1000} = \dfrac{1}{8}$

17. $2.25\% = \dfrac{2.25}{100} = \dfrac{225}{10\,000} = \dfrac{9}{400}$

19. $\dfrac{1}{8}\% = \dfrac{1}{8(100)} = \dfrac{1}{800}$

21. $\dfrac{3}{4}\% = \dfrac{3}{4(100)} = \dfrac{3}{400}$

23. $6.25\% = \dfrac{6.25}{100} = \dfrac{625}{10\,000} = \dfrac{1}{16}$

25. $16\dfrac{2}{3}\% = \dfrac{50}{3}\% = \dfrac{50}{3(100)} = \dfrac{1}{6}$

27. $0.75\% = \dfrac{0.75}{100} = \dfrac{75}{10\,000} = \dfrac{3}{400}$

29. $0.1\% = \dfrac{0.1}{100} = \dfrac{1}{1000}$

31. $83\dfrac{1}{3}\% = \dfrac{250}{3}\% = \dfrac{250}{3(100)} = \dfrac{5}{6}$

33. $133\dfrac{1}{3}\% = \dfrac{400}{3}\% = \dfrac{400}{3(100)} = \dfrac{4}{3}$

35. $166\dfrac{2}{3}\% = \dfrac{500}{3}\% = \dfrac{500}{3(100)} = \dfrac{5}{3}$

C. 1. $3.5 = 3.5(100\%) = 350\%$

3. $0.005 = 0.005(100\%) = 0.5\%$

5. $0.025 = 0.025(100\%) = 2.5\%$

7. $0.125 = 0.125(100\%) = 12.5\%$

9. $0.225 = 0.225(100\%) = 22.5\%$

11. $1.45 = 1.45(100\%) = 145\%$

13. $0.0025 = 0.0025(100\%) = 0.25\%$

15. $0.09 = 0.09(100\%) = 9\%$

17. $\dfrac{3}{4} = 0.75(100\%) = 75\%$

19. $\dfrac{5}{3} = 1.\dot{6} = 1.6(100\%) = 166.\dot{6}\%$

21. $\dfrac{9}{200} = 0.045(100\%) = 4.5\%$

23. $\dfrac{3}{400} = 0.0075(100\%) = 0.75\%$

25. $\dfrac{9}{800} = 0.01125(100\%) = 1.125\%$

27. $\dfrac{3}{8} = 0.375(100\%) = 37.5\%$

29. $\dfrac{4}{3} = 1.\dot{3}(100\%) = 133.\dot{3}\%$

31. $\dfrac{13}{20} = 0.65(100\%) = 65\%$

Exercise 1.4

A. 1. Total hectares $= 3\dfrac{3}{4} + 2\dfrac{2}{3} + 3\dfrac{5}{8} + 4\dfrac{5}{6}$

$= 3.75 + 2.6666667 + 3.625 + 4.8333333$

$= 14.875$

Total sales value $= 14.875 \times 27\,500 = \$409\,062.50$

3. Assessed value $= \dfrac{6}{11} \times 56\,100 = 6 \times 5100 = \$30\,600$

Property tax $= 30\,600 \times \dfrac{3.75}{100} = \$1147.50$

5.
$$
\begin{array}{lll}
64 \times 0.75 & = & \$\ 48.00 \\
54 \times 0.83\dfrac{1}{3} = 54 \times 0.8333333 & = & 45.00 \\
72 \times 0.375 & = & 27.00 \\
42 \times 1.33 = 42 \times 1.3333333 & = & \underline{\ \ 56.00} \\
& & \$176.00
\end{array}
$$

B. 1.
$$
\begin{array}{ll}
1100 \times 0.385 & = \$\ 423.50 \\
1600 \times 0.415 & = \ \ 664.00 \\
1400 \times 0.425 & = \ \ \underline{595.00} \\
\text{Total cost} & = \$1682.50
\end{array}
$$

Average cost per litre $= \dfrac{1682.50}{4100} = 0.4103659 = \$0.41$

3. Weighted hours $= 3 \times 4 + 5 \times 2 + 2 \times 6 + 4 \times 2 + 4 \times 1 + 2 \times 6$

$= 12 + 10 + 12 + 8 + 4 + 12$

$= 58$

Total hours $= 3 + 5 + 2 + 4 + 4 + 2 = 20$

Grade-point average $= \dfrac{58}{20} = 2.9$

5. (a) Simple average of unit prices

$$= \frac{10.00 + 10.60 + 11.25 + 9.50 + 9.20 + 12.15}{6} = \frac{62.70}{6} = \$10.45$$

(b) Number of units purchased $= \dfrac{\text{Amount invested}}{\text{Unit price}}$

Date	Amount Invested	Unit Price	Number of Units Purchased
February 1	200.00	10.00	$\dfrac{200.00}{10.00} = 20.000$
March 1	200.00	10.60	$\dfrac{200.00}{10.60} = 18.868$
April 1	200.00	11.25	$\dfrac{200.00}{11.25} = 17.778$
May 1	200.00	9.50	$\dfrac{200.00}{9.50} = 21.053$
June 1	200.00	9.20	$\dfrac{200.00}{9.20} = 21.739$
July 1	200.00	12.15	$\dfrac{200.00}{12.15} = 16.461$
		Total number of units purchased	115.899

(c) Average cost of units purchased $= \dfrac{1200.00}{115.899} = \$10.35$

(d) Value on July 31 $= 115.899(11.90) = \$1379.20$

Exercise 1.5

A. 1. (a) Annual salary $= \$22\,932.00$

Semimonthly payment $= \dfrac{22\,932.00}{24} = \$955.50$

(b) Weekly pay $= \dfrac{22\,932.00}{52} = \$441.00$

Hourly rate $= \dfrac{441.00}{36} = \$12.25$

(c)
Regular pay	$=$	$\$\ 955.50$
Overtime pay $= 11 \times 12.25 \times 1.5$	$=$	$\underline{\ \ \ 202.13}$
Gross pay	$=$	$\$1157.63$

3.   (a) Monthly pay = \$1101.10
            Yearly pay = 1101.10 × 12 = \$13 213.20
            Weekly pay = 13 213.20 ÷ 52 = \$254.10
            Hourly rate of pay = 254.10 ÷ 35 = \$7.26

      (b) Regular pay for May   = \$1101.10
            Overtime pay = 7.75 × 7.26 × 1.5 =   84.40
            Gross pay   = \$1185.50

5.   (a) Biweekly payment = \$1123.00
            Annual salary = 1123.00 × 22 = \$24 706
            Daily pay = 24 706 ÷ 200 = \$123.53
            Hourly rate = 123.53 ÷ 7.5 = \$16.47

      (b) Regular pay   = \$1123.00
            Less: two days = 123.53 × 2 =   247.06
            Gross pay   = \$ 875.94

7.   Net sales = \$16 244.00

Commission: $8\frac{1}{4}$%on first \$6000.00   = \$ 495.00

                $9\frac{3}{4}$%on next \$6000.00   =   585.00

                11.5% on \$(16 244.00 − 12 000.00) =   488.06
      Total commission   = \$1568.06

9.   (a) Sales = \$4125.00
            Base salary on quota of \$4500 = \$225.00

      (b) Sales = \$6150.00
            Base salary on quota of \$4500   = \$225.00

            Commission = $6\frac{1}{2}$% on \$1650 = 0.065 × \$1650 =   107.25
            Gross earnings   = \$332.25

11.   Gross sales   = \$21 440.00
      Less: returns = 5% of \$21 440.00 =   1 072.00
        Net sales   = \$20 368.00

      Rate of commission = $\dfrac{1\ 884.04}{20\ 368.00}$ = 9.25%

13.   Net sales = $\dfrac{\$\ Commission}{Rate}$ = $\dfrac{\$2036.88}{0.1125}$ = \$18 105.60

      Net sales   =   gross sales − returns
     18 105.60 = $S − 0.08S$
        0.92S =   18 105.60
           S =   19 680
  Gross sales were \$19 680.

15. **Method A**

Regular earnings $= 40 \times 8.42$	$= \$336.80$
Overtime pay $= 7 \times 8.42 \times 1.5$	$= \underline{88.41}$
Total earnings	$= \$425.21$

Method B

Earnings at regular rate: 47×8.42	$= \$395.74$
Overtime premium: $7 \times 8.42 \times 0.5$	$= \underline{29.47}$
Gross earnings	$= \$425.21$

17. Rate of pay $= \dfrac{319.44}{44} = \$7.26$

Exercise 1.6

1.

Month	GST collected 7% of sales	GST paid 7% of purchases	GST payable (GST receivable)
January	$ 38 283.00	$ 10 348.24	$27 934.76
February	17 080.00	4 865.00	12 215.00
March	41 160.00	60 620.00	(19 460.00)
April	45 521.00	31 500.00	14 021.00
May	10 976.00	6 866.86	4 109.14
5-month totals	$153 020.00	$114 200.10	$38 819.90

Cook's owes the government $38 819.90.

3. Savings on GST $= 7\%$ of $\$15.00 = 0.07(15.00) = \$1.05$

5. At Blackcomb, B.C.

Cost of ski pass	$= \$84.00$
GST $= 7\%$ of $\$84.00 = 0.07(84.00)$	5.88
PST $= 7\%$ of $\$84.00 = 0.07(84.00)$	$= \underline{5.88}$
Amount paid at Blackcomb	$= \$95.76$

At Mont Tremblant, Que.

Cost of ski pass		$= \$84.00$
GST $= 7\%$ of $\$84.00 = 0.07(84.00)$		5.88
PST $= 8\%$ of $\$84.00 = 0.08(84.00)$	$=\$6.72$	
$+\ 8\%$ of $\$5.88 = 0.08(5.88)$	$= \underline{0.47}$	$\underline{7.19}$
Amount paid at Mont Tremblant		$= \underline{\$97.07}$
Difference $= 97.07 - 95.76$		$\$ 1.31$
made up by 1% in PST $= 1\%$ of $\$84.00 =$	$\$0.84$	
and 8% on GST $= 8\%$ of $\$5.88 \quad =$	0.47	

7. Property tax $= 125\,000\left(\dfrac{22.751}{1000}\right) = \$2843.88$

9. Semi-annual tax rate $= \dfrac{1420.79}{196\,000.00} = 0.00724893$

    Semi-annual mill rate $= 0.00724893(1000) = 7.24893$

    The annual mill rate $= 2(7.24893) = 14.49786$

Review Exercise

1. (a) $32 - 24 \div 8 = 32 - 3 = 29$

    (b) $(48 - 18) \div 15 - 10 = 30 \div 15 - 10 = 2 - 10 = -8$

    (c) $(8 \times 6 - 4) \div (16 - 4 \times 3) = (48 - 4) \div (16 - 12) = 44 \div 4 = 11$

    (d) $9(6 - 2) - 4(3 + 4) = 9(4) - 4(7) = 36 - 28 = 8$

    (e) $\dfrac{108}{0.18 \times \frac{216}{360}} = \dfrac{108}{0.18 \times 0.6} = \dfrac{108}{0.108} = 1000$

    (f) $\dfrac{288}{2400 \times \frac{292}{365}} = \dfrac{288}{2400 \times 0.8} = \dfrac{288}{1920} = 0.15$

    (g) $320\left(1 + 0.10 \times \dfrac{225}{360}\right) = 320(1 + 0.0625) = 320(1.0625) = 340$

    (h) $1000\left(1 - 0.12 \times \dfrac{150}{360}\right) = 1000(1 - 0.05) = 1000(0.95) = 950$

    (i) $\dfrac{660}{1 + 0.14 \times \frac{144}{360}} = \dfrac{660}{1 + 0.056} = \dfrac{660}{1.056} = 625$

    (j) $\dfrac{1120}{1 - 0.13 \times \frac{292}{365}} = \dfrac{1120}{1 - 0.104} = \dfrac{1120}{0.896} = 1250$

3. (a) $50\% = \dfrac{50}{100} = \dfrac{1}{2}$

    (b) $37\frac{1}{2}\% = \dfrac{37.5}{100} = \dfrac{375}{1000} = \dfrac{3}{8}$

    (c) $16\frac{2}{3}\% = \dfrac{16\frac{2}{3}}{100} = \dfrac{\frac{50}{3}}{\frac{100}{1}} = \dfrac{1}{6}$

    (d) $166\frac{2}{3}\% = \dfrac{100 + 66\frac{2}{3}}{100} = 1 + \dfrac{2}{3} = \dfrac{5}{3}$

(e) $\dfrac{1}{2}\% = \dfrac{\frac{1}{2}}{100} = \dfrac{1}{2} \times \dfrac{1}{100} = \dfrac{1}{200}$

(f) $7.5\% = \dfrac{7.5}{100} = \dfrac{75}{1000} = \dfrac{3}{40}$

(g) $0.75\% = \dfrac{3}{4}\% = \dfrac{3}{400}$

(h) $\dfrac{5}{8}\% = \dfrac{5}{800} = \dfrac{1}{160}$

5. (a) $4\dfrac{1}{3} + 3\dfrac{3}{4} + 5\dfrac{1}{2} + 6\dfrac{5}{8}$
$= 4.3333333 + 3.75 + 5.5 + 6.625 = 20.2083333 \text{ kg}$

(b) $20.2083333 \times 1.20 = \$24.25$

(c) $20.2083333 \div 4 = 5.0520833 = 5.05 \text{ kg}$

(d) $24.25 \div 4 = 6.0625 = \$6.06$

7. (a) $\dfrac{15.45 + 12.20 + 9.60 + 7.50}{4} = \dfrac{44.75}{4} = 11.1875 = \$11.19$

(b)
$$
\begin{array}{rcl}
15.45 \times 2 & = & \$\ 30.90 \\
12.20 \times 6 & = & 73.20 \\
9.60 \times 9 & = & 86.40 \\
7.50 \times \underline{13} & = & \underline{97.50} \\
30 & & \$288.00
\end{array}
$$
Average rate $= \dfrac{288}{30} = \$9.60$

9.
$$
\begin{array}{lll}
\text{January 1 – March 31:} & 12\,000 \times 3 & = \$\ 36\,000 \\
\text{April 1 – May 31:} & 14\,400 \times 2 & = \ 28\,800 \\
\text{June 1 – September 30:} & 12\,960 \times 4 & = \ 51\,840 \\
\text{October 1 – December 31:} & 15\,840 \times \underline{\ 3} & = \ \underline{47\,520} \\
\quad \text{Total} & \qquad\quad 12 & \$164\,160
\end{array}
$$
Average monthly investment $= \dfrac{\$164\,160}{12} = \$13\,680$

11. (a) Monthly remuneration $= \dfrac{17\ 472.00}{12} = \$1456.00$

(b) Weekly pay $= 17\ 472.00 \div 52 = \$336.00$
Hourly rate $= 336.00 \div 35 = \$9.60$

(c) Gross pay for month $= \$1693.60$
Regular gross pay $= \underline{\ \ 1456.00}$
Overtime pay $= \$\ 237.60$
Overtime hours $= 237.60 \div (9.60 \times 1.5) = 16.5$

13. (a) Gross sales $= \$5580.00$
Less: returns $= \underline{\ \ \ \ 60.00}$
Net sales $= \$5520.00$
Commission: 4% of \$3000.00 $= \$120.00$
8% of \$1500.00 $= \ \ 120.00$
12.5% of \$[5520 − 4500] $= \underline{\ \ 127.50}$
Gross earnings $= \$367.50$

(b) Average hourly rate $= 367.50 \div 43 = \$8.55$

15. Gross sales $= \$21\ 500.00$
Less: returns $= \underline{\ \ \ \ 325.00}$
Net sales $= \$21\ 175.00$
Commission: 7.5% of \$7000.00 $= \$\ 525.00$
9% of \$8000.00 $= \ \ 720.00$
11% of \$6175.00 $= \underline{\ \ 679.25}$
Gross earnings $= \$1924.25$

17. Gross earnings $= \$321.30$
Base salary $= \underline{\ \ 255.00}$
Commission $= \$\ 66.30$
Commission sales $= 6560.00 − 5000 = \$1560.00$
Rate of commission $= 66.30 \div 1560.00 = 0.0425 = 4.25\%$

19. Hours worked $= 47$
Regular hours $= \underline{40}$
Overtime hours $= 7$
7 overtime hours are equivalent to $7 \times 1.5 = 10.5$ regular hours.
Total hours paid at regular rate $= 40 + 10.5 = 50.5$
Hourly rate of pay $= \dfrac{426.22}{50.5} = \$8.44$

21. Gross earnings $= \$328.54$
Less: base salary $= \underline{\ \ 280.00}$
Commission $= \$\ 48.54$
Commission sales $= 48.54 \div 0.06 = \$809.00$
Net sales $= 5000.00 + 809.00 = \$5809.00$
Gross sales $= 5809.00 + 136.00 = \$5945.00$

23. Gross earnings   = $349.05
    Regular earnings = 35 × 7.80   =   273.00
        Overtime pay   = $ 76.05
    Overtime hours = 76.05 ÷ (7.80 × 1.5) = 6.5
    Number of hours worked = 35 + 6.5 = 41.5

25. GST collected:
    Parts: 7% of $ 75 000
    Labour: 7% of $ 65 650
    Total   7% of $140 650 = 0.07(140 650)   = $9845.50

    GST paid
    Parking fees: 7% of $ 4 000
    Supplies: 7% of $55 000
    Utilities: 7% of $ 2 000
    Other: 7% of $ 3 300
    Total   7% of $64 300 = 0.07(64 300)   =   4501.00
           GST remittance $5344.50

27. Ticket price $100.00
    Ontario PST = 8% of 100.00   8.00
        Amount paid $108.00

    Net price + 7% of net price   =   Ticket price
                N + 0.07N   =   100.00
                     1.07N   =   100.00

$$\text{Net price} = \frac{100.00}{1.07} = \$93.46$$

    Total tax paid = 108.00 − 93.46 = $14.54

$$\text{Tax rate based on net price} = \frac{14.54}{93.46} = 15.56\%$$

29. (a) Mill rate $= \dfrac{45\,567\,000}{975\,500\,000}(1000) = 46.71143$

    (b) Property tax $= 35\,000\left(\dfrac{46.71143}{1000}\right) = \$1634.90$

    (c) Increase in mill rate $= \dfrac{2\,000\,000}{975\,500\,000}(1000) = 2.05023$

    (d) Additional property tax $= 35\,000\left(\dfrac{2.05023}{1000}\right) = \$71.76$

1. (a) $4320\left(1 + 0.18 \times \dfrac{45}{360}\right) = 4320(1 + 0.0225) = 4417.2$

    (b) $2160\left(0.15 \times \dfrac{105}{360}\right) = 2160(0.04375) = 94.5$

    (c) $2880\left(1 - 0.12 \times \dfrac{285}{360}\right) = 2880(1 - 0.095) = 2606.4$

    (d) $\dfrac{410.40}{0.24 \times \frac{135}{360}} = \dfrac{410.40}{0.09} = 4560$

    (e) $\dfrac{5124}{1 + 0.09 \times \frac{270}{360}} = \dfrac{5124}{1 + 0.0675} = 4800$

3. (a) $2\dfrac{1}{2}\% = \dfrac{5}{2}\% = \dfrac{5}{2} \times \dfrac{1}{100} = \dfrac{5}{200} = \dfrac{1}{40}$

    (b) $116\dfrac{2}{3}\% = 100\% + 16\dfrac{2}{3}\% = 1 + \dfrac{16\frac{2}{3}}{100} = 1 + \dfrac{\frac{50}{3}}{100} = 1 + \dfrac{50}{300}$
    $= 1 + \dfrac{1}{6} = \dfrac{7}{6}$

5.
January 1 – February 28:	7200	× 2	= $14 400
March 1 – July 31:	6720	× 5	= 33 600
August 1 – September 30:	7320	× 2	= 14 640
October 1 – December 31:	7440	× 3	= 22 320
Total		12	$84 960

Average monthly balance $= \dfrac{84\,960}{12} = \$7080$

7.
5 × 9	=	$ 45
6 × 7	=	42
3 × 8	=	24
6 × 6	=	36
20 Total	=	$147

Average price $= \dfrac{147}{20} = \$7.35$

9. Assessed value $= \dfrac{2}{13} \times 130\,000 = \$20\,000$

    Property tax $= \$20\,000 \times \dfrac{3.25}{100} = \$650$

11. Weekly pay = 26 478.40 ÷ 52 = $509.20
     Hourly pay = 509.20 ÷ 38 = $13.40
     Regular monthly pay = 26 478.40 ÷ 12   = $2206.53
     Overtime earnings = 13.40 × 8.75 × 1.5 = <u>175.88</u>
       Gross pay   = $2382.41

13. Total hours   = 52.5
     Regular hours   = <u>42.0</u>
       Overtime hours = 10.5
     At time-and-a-half, 10.5 overtime hours are equivalent to
     10.5 × 1.5 = 15.75 regular hours
     Hourly rate of pay = $\dfrac{513.98}{57.75}$ = $8.90

15. Annual salary = 780.00 × 24 = $18 720.00
     Weekly pay = 18 720.00 ÷ 52 = $360.00
     Hourly rate of pay = 360.00 ÷ 40 = $9.00

17. Property Tax = Assessed Value × Mill Rate

     2502.50 = Assessed Value × $\dfrac{55}{1000}$

     Assessed Value = $\dfrac{2502.50(1000)}{55}$ = $45 500.00

2   Review of basic algebra

Exercise 2.1

A.   1. $19a$

3. $-a - 10$

5. $0.8x$

7. $1.4x$

9. $-x^2 - x - 8$

11. $2x - 3y - x - 4y = x - 7y$

13. $a^2 - ab + b^2 - 3a^2 - 5ab + 4b^2 = -2a^2 - 6ab + 5b^2$

15. $6 - 4x + 3y - 1 - 5x - 2y + 9 = 14 - 9x + y$

B.   1. $-12x$

3. $-10ax$

5. $-2x^2$

7. $60xy$

9. $-2x + 4y$

11. $2ax^2 - 3ax - a$

13. $20x - 24 - 6 + 15x = 35x - 30$

15. $-15ax + 3a + 5a - 2ax - 3ax - 3a = -20ax + 5a$

17. $3x^2 - x + 6x - 2 = 3x^2 + 5x - 2$

19. $x^3 - x^2y + xy^2 + x^2y - xy^2 + y^3 = x^3 + y^3$

21. $10x^2 - 8x - 5x + 4 - 3x^2 + 21x - 5x + 35 = 7x^2 + 3x + 39$

C. 1. $4ab$

3. $4x$

5. $10m - 4$

7. $-2x^2 + 3x + 6$

D. 1. $3x - 2y - 3 = 3(-4) - 2(-5) - 3 = -12 + 10 - 3 = -5$

3. $\dfrac{RP(n + 1)}{2N} = \dfrac{0.21 \times 1200 \times (77 + 1)}{2 \times 26} = 378$

5. $\dfrac{I}{RT} = \dfrac{198}{0.165 \times \frac{146}{365}} = \dfrac{198}{0.165 \times 0.40} = 3000$

7. $P(1 + RT) = 880\left(1 + 0.12 \times \dfrac{75}{360}\right)$
$= 880(1 + 0.025) = 880(1.025) = 902$

9. $\dfrac{P}{1 - dt} = \dfrac{1253}{1 - 0.135 \times \frac{280}{360}} = \dfrac{1253}{1 - 0.105} = \dfrac{1253}{0.895} = 1400$

Exercise 2.2

A. 1. 81

3. 16

5. $\dfrac{16}{81}$

7.   0.25

9.   1

11. $\dfrac{1}{9}$

13.   125

15. $\dfrac{1}{1.01}$

B. 1. $2^5 \times 2^3 = 2^{5+3} = 2^8$

3. $4^7 \div 4^4 = 4^{7-4} = 4^3$

5. $(2^3)^5 = 2^{3\times5} = 2^{15}$

7. $a^4 \times a^{10} = a^{4+10} = a^{14}$

9. $3^4 \times 3^6 \times 3 = 3^{4+6+1} = 3^{11}$

11. $\dfrac{6^7 \times 6^3}{6^9} = 6^{7+3-9} = 6$

13. $\left(\dfrac{3}{5}\right)^4\left(\dfrac{3}{5}\right)^7 = \left(\dfrac{3}{5}\right)^{4+7} = \dfrac{3^{11}}{5^{11}}$

15. $\left(-\dfrac{3}{2}\right)\left(-\dfrac{3}{2}\right)^6\left(-\dfrac{3}{2}\right)^4 = \left(-\dfrac{3}{2}\right)^{1+6+4} = \dfrac{(-3)^{11}}{2^{11}}$

17. $(1.025)^{80}(1.025)^{70} = (1.025)^{80+70} = 1.025^{150}$

19. $\left[1.04^{20}\right]^4 = 1.04^{20\times4} = 1.04^{80}$

21. $(1 + i)^{100}(1 + i)^{100} = (1 + i)^{100+100} = (1 + i)^{200}$

23. $\left[(1 + i)^{80}\right]^2 = (1 + i)^{80 \times 2} = (1 + i)^{160}$

25. $(ab)^5 = a^5 b^5$

27. $(m^3 n)^8 = m^{24} n^8$

29. $2^3 \times 2^5 \times 2^{-4} = 2^{3+5-4} = 2^4 = 16$

31. $\left(\dfrac{a}{b}\right)^{-8} = \dfrac{b^8}{a^8}$

Exercise 2.3

A. 1. $\sqrt{5184} = 72$

3. $\sqrt[7]{2187} = 3$

5. $\sqrt[20]{4.3184} = 1.0758857$

7. $\sqrt[6]{1.0825} = 1.0132999$

B. 1. $3025^{\frac{1}{2}} = 55$

3. $525.21875^{\frac{2}{5}} = 12.25$

5. $\sqrt[12]{1.125^7} = 1.0711221$

7. $4^{\left(-\frac{1}{3}\right)} = \dfrac{1}{4^{\frac{1}{3}}} = \dfrac{1}{1.5874011} = 0.6299605$

9. $\dfrac{1.03^{60} - 1}{0.03} = \dfrac{5.891603 - 1}{0.03} = 163.0534333$

Exercise 2.4

A. 1. $2^9 = 512$
   $9 = \log_2 512$

3. $5^{-3} = \dfrac{1}{125}$
   $-3 = \log_5 \dfrac{1}{125}$

5. $e^{2j} = 18$
   $2j = \log_e 18$
   or $2j = \ln 18$

B. 1. $\log_2 32 = 5$
   $2^5 = 32$

3. $\log_{10} 10 = 1$
   $10^1 = 10$

C. 1. $\ln 2 = 0.6931472$

3. $\ln 0.105 = -2.2537949$

5. $\ln\left[\dfrac{2000}{1.09^9}\right]$ $= \ln 2000 - \ln 1.09^9$

   $= \ln 2000 - 9(\ln 1.09)$
   $= 7.6009025 - 9(0.0861777)$
   $= 7.6009025 - 0.7755993$
   $= 6.8253032$

Exercise 2.5

A. 1. $15x = 45$
   $x = 3$

3. $0.9x = 72$
   $x = 80$

5. $\dfrac{1}{6}x = 3$
   $x = 18$

7. $\dfrac{3}{5}x = -21$
   $\dfrac{1}{5}x = -7$
   $x = -35$

9. $x - 3 = -7$
   $x = -4$

11. $x + 6 = -2$
    $x = -8$

13. $4 - x = 9 - 2x$
    $x = 5$

15. $x + 0.6x = 32$
    $1.6x = 32$
    $x = 20$

17. $x - 0.04x = 192$
    $0.96x = 192$
    $x = 200$

B. 1. $3x + 5 = 7x - 11$
   $-4x = -16$
   $x = 4$

   LS: $3x + 5 = 3(4) + 5$
   $= 12 + 5$
   $= 17$

   RS: $7x - 11 = 7(4) - 11$
   $= 28 - 11$
   $= 17$

3.
$$2 - 3x - 9 = 2x - 7 + 3x$$
$$-3x - 7 = 5x - 7$$
$$-8x = 0$$
$$x = 0$$

LS: $= 2 - 3x - 9$
$= 2 - 3(0) - 9$
$= -7$

RS: $= 2x - 7 + 3x$
$= 2(0) - 7 + 3(0)$
$= -7$

Exercise 2.6

A. 1.
$$12x - 4(9x - 20) = 320$$
$$12x - 36x + 80 = 320$$
$$-24x = 240$$
$$x = -10$$

LS $= 12(-10) - 4[9(-10) - 20]$
$= -120 - 4[-90 - 20]$
$= -120 + 440$
$= 320$
RS $= 320$

3.
$$3(2x - 5) - 2(2x - 3) = -15$$
$$6x - 15 - 4x + 6 = -15$$
$$2x - 9 = -15$$
$$2x = -6$$
$$x = -3$$

LS $= 3[2(-3) - 5] - 2[2(-3) - 3]$
$= 3[-6 - 5] - 2[-6 - 3]$
$= 3(-11) - 2(-9)$
$= -33 + 18$
$= -15$
RS $= -15$

B. 1.
$$x - \frac{1}{4}x = 15$$
$$4x - x = 60$$
$$3x = 60$$
$$x = 20$$

3. $\dfrac{2}{3}x - \dfrac{1}{4} = -\dfrac{7}{4} - \dfrac{5}{6}x$

$8x - 3 = -21 - 10x$

$18x = -18$

$x = -1$

5. $\dfrac{3}{4}x + 4 = \dfrac{113}{24} - \dfrac{2}{3}x$

$18x + 96 = 113 - 16x$

$34x = 17$

$x = \dfrac{1}{2}$

C. 1. $\dfrac{3}{4}(2x - 1) - \dfrac{1}{3}(5 - 2x) = -\dfrac{55}{12}$

$9(2x - 1) - 4(5 - 2x) = -55$

$18x - 9 - 20 + 8x = -55$

$26x - 29 = -55$

$26x = -26$

$x = -1$

3. $\dfrac{2}{3}(2x - 1) - \dfrac{3}{4}(3 - 2x) = 2x - \dfrac{20}{9}$

$24(2x - 1) - 27(3 - 2x) = 72x - 80$

$48x - 24 - 81 + 54x = 72x - 80$

$102x - 105 = 72x - 80$

$30x = 25$

$x = \dfrac{5}{6}$

D. 1. $A = \dfrac{1}{2}bh$

$2A = bh$

$h = \dfrac{2A}{b}$

3. $F = \dfrac{9}{5}C + 32$

$5F = 9C + 160$

$-9C = -5F + 160$

$9C = 5F - 160$

$C = \dfrac{5F - 160}{9}$

$C = \dfrac{5}{9}(F - 32)$

24

5. $A = P(1 + rt)$

$$\frac{A}{P} = 1 + rt$$

$$\frac{A}{P} - 1 = rt$$

$$r = \frac{\dfrac{A}{P} - 1}{t}$$

$$r = \frac{\dfrac{A - P}{P}}{t}$$

$$r = \frac{A - P}{Pt}$$

Exercise 2.7

A.   1.   Let the cost be $\$x$.

Selling price $= \$\left(x + \dfrac{3}{4}x\right)$

$\therefore\ x + \dfrac{3}{4}x = 49.49$

$4x + 3x = 197.96$

$7x = 197.96$

$x = 28.28$

The cost was $28.28.

3.   Let the last month's index be x.

This month's index $= x - \dfrac{1}{12}x$

$\therefore\ x - \dfrac{1}{12}x = 176$

$12x - x = 2112$

$11x = 2112$

$x = 192$

Last month the index was 192.

5.   Let Vera's sales be $\$x$.

Nancy's sales $= \$(3x - 140)$

Total sales $= \$(x + 3x - 140)$

$\therefore\ x + 3x - 140 = 940$

$4x = 1080$

$x = 270$

Nancy's sales $= 3(270) - 140 = \$670$.

7.   Let Ken's investment be $x.

Marina's investment $= \$\left(\dfrac{2}{3}x + 2500\right)$

Total investment $= \$\left(x + \dfrac{2}{3}x + 2500\right)$

$\therefore x + \dfrac{2}{3}x + 2500 = 55\,000$

$\qquad\qquad \dfrac{5x}{3} = 52\,500$

$\qquad\qquad\quad x = 31\,500$

Martina's investment is $\dfrac{2}{3} \times 31\,500 + 2500 = \$23\,500$.

9.   Let the number of type A lights be x.
Number of type B lights $= 60 - x$.
Value of type A lights $= \$40x$.
Value of type B lights $= \$(60 - x)50$.
$\therefore 40x + 50(60 - x) = 2580$
$\quad 40x + 3000 - 50x = 2580$
$\qquad\qquad\quad -10x = -420$
$\qquad\qquad\qquad\; x = 42$
The number of type B lights is 18.

11.   Let the number of dimes be x.
Number of nickels $= 3x - 4$

Number of quarters $= \dfrac{3}{4}x + 1$

Value of the dimes $= 10x$ cents
Number of nickels $= 5(3x - 4)$ cents

Value of quarters $= 25\left(\dfrac{3}{4}x + 1\right)$ cents

$\therefore 10x + 5(3x - 4) + 25\left(\dfrac{3}{4}x + 1\right) = 880$

$\qquad 10x + 15x - 20 + \dfrac{75}{4}x + 25 = 880$

$\qquad\qquad\qquad\quad 25x + \dfrac{75}{4}x = 875$

$\qquad\qquad\qquad\qquad\quad 175x = 3500$

$\qquad\qquad\qquad\qquad\qquad x = 20$

Bruce has 20 dimes, 56 nickels, and 16 quarters.

Review Exercise

1.   (a) $3x - 4y - 3y - 5x = -2x - 7y$

(b) $2x - 0.03x = 1.97x$

(c) $(5a - 4) - (3 - a) = 5a - 4 - 3 + a = 6a - 7$

(d) $-(2x - 3y) - (-4x + y) + (y - x) = -2x + 3y + 4x - y + y - x = x + 3y$

(e) $\left(5a^2 - 2b - c\right) - \left(3c + 2b - 4a^2\right)$
$= 5a^2 - 2b - c - 3c - 2b + 4a^2 = 9a^2 - 4b - 4c$

(f) $-(2x - 3) - \left(x^2 - 5x + 2\right) = -2x + 3 - x^2 + 5x - 2 = -x^2 + 3x + 1$

3.   (a)   for $x = -2$, $y = 5$,
$3xy - 4x - 5y = 3(-2)(5) - 4(-2) - 5(5) = -30 + 8 - 25 = -47$

(b)   for $a = -\dfrac{1}{4}$, $b = \dfrac{2}{3}$,
$-5(2a - 3b) - 2(a + 5b)$
$= -10a + 15b - 2a - 10b$
$= -12a + 5b$
$= -12\left(-\dfrac{1}{4}\right) + 5\left(\dfrac{2}{3}\right) = 3 + 3\dfrac{1}{3} = 6\dfrac{1}{3}$

(c)   for N = 12, C = 432, P = 1800, $n = 35$,
$$\frac{2NC}{P(n + 1)} = \frac{2 \times \overset{1}{12} \times \overset{48}{432}}{\underset{\underset{100}{900}}{1800} \times (35 + 1)} = \frac{\overset{1}{12} \times \overset{16}{48}}{100 \times \underset{\underset{1}{3}}{36}} = \frac{16}{100} = 0.16$$

(d)   for I = 600, R = 0.15, P = 7300,
$$\frac{365\,I}{RP} = \frac{\overset{1}{365} \times \overset{\overset{30}{2}}{600}}{\underset{\underset{1}{20}}{0.15} \times \underset{0.01}{7300}} = \frac{2}{0.01} = 200$$

(e)   for A = 720, d = 0.135, $t = \dfrac{280}{360}$,

$A(1 - dt) = 720\left(1 - 0.135 \times \dfrac{\overset{7}{280}}{\underset{9}{360}}\right) = 720(1 - 0.015 \times 7)$
$= 720(1 - 0.105) = 720(0.895) = 644.40$

(f)   for S = 2755, R = 0.17, T = $\dfrac{219}{365}$,

$$\frac{S}{1 + RT} = \frac{2755}{1 + 0.17 \times \frac{219}{365}} = \frac{2755}{1 + 0.034 \times 3} = \frac{2755}{1 + 0.102} = 2500$$

5.   (a) $\sqrt{0.9216} = 0.96$

      (b) $\sqrt[6]{1.075} = 1.0121264$

      (c) $14.974458^{1/40} = 1.07$

      (d) $1.08^{-5/12} = \dfrac{1}{1.08^{5/12}} = 0.9684416$

      (e) $\ln 3 = 1.0986123$

      (f) $\ln 0.05 = -2.9957323$

      (g) $\ln\left(\dfrac{5500}{1.10^{16}}\right) = \ln 5500 - \ln 1.10^{16}$

$$= \ln 5500 - 16 \ln 1.10$$
$$= 8.6125034 - 16(0.0953102)$$
$$= 8.6125034 - 1.5249632$$
$$= 7.0875402$$

      (h) $\ln\left[375(1.01)\left(\dfrac{1 - 1.01^{-72}}{0.01}\right)\right] = \ln 375 + \ln 1.01 + \ln\left(1 - 1.01^{-72}\right) - \ln 0.01$

$$= \ln 375 + \ln 1.01 + \ln (1 - 0.4884961) - \ln 0.01$$
$$= \ln 375 + \ln 1.01 + \ln 0.5115004 - \ln 0.01$$
$$= 5.9269260 + 0.0099503 - 0.6704069 - (-4.6051702)$$
$$= 9.8716396$$

7.   (a) $-9(3x - 8) - 8(9 - 7x) = 5 + 4(9x + 11)$
$$-27x + 72 - 72 + 56x = 5 + 36x + 44$$
$$29x = 49 + 36x$$
$$-7x = 49$$
$$x = -7$$

Check   LS $= -9[3(-7) - 8] - 8[9 - 7(-7)] = -9(-29) - 8(58) = -203$

RS $= 5 + 4[9(-7) + 11] = 5 + 4(-52) = 5 - 208 = -203$

(b) $\quad 21x - 4 - 7(5x - 6) = 8x - 4(5x - 7)$

$$21x - 4 - 35x + 42 = 8x - 20x + 28$$
$$-14x + 38 = -12x + 28$$
$$-2x = -10$$
$$x = 5$$

Check $\text{LS} = 21(5) - 4 - 7[5(5) - 6] = 105 - 4 - 7(19) = 101 - 133 = -32$

$\quad\quad\text{RS} = 8(5) - 4[5(5) - 7] = 40 - 4(18) = 40 - 72 = -32$

(c)
$$\frac{5}{7}x + \frac{1}{2} = \frac{5}{14} + \frac{2}{3}x$$
$$42\left(\frac{5}{7}x\right) + 42\left(\frac{1}{2}\right) = 42\left(\frac{5}{14}\right) + 42\left(\frac{2}{3}x\right)$$
$$6(5x) + 21(1) = 3(5) + 14(2x)$$
$$30x + 21 = 15 + 28x$$
$$2x = -6$$
$$x = -3$$

Check $\text{LS} = \dfrac{5}{7}(-3) + \dfrac{1}{2} = \dfrac{-30 + 7}{14} = -\dfrac{23}{14}$

$\quad\quad\text{RS} = \dfrac{5}{14} + \dfrac{2}{3}(-3) = \dfrac{5}{14} - 2 = -\dfrac{23}{14}$

(d)
$$\frac{4x}{3} + 3 = \frac{9}{8} - \frac{x}{6}$$
$$8(4x) + 24(2) = 3(9) - 4(x)$$
$$32x + 48 = 27 - 4x$$
$$36x = -21$$
$$x = -\frac{7}{12}$$

Check $\text{LS} = \dfrac{4}{3}\left(-\dfrac{7}{12}\right) + 2 = -\dfrac{28}{36} + 2 = -\dfrac{7}{9} + \dfrac{18}{9} = \dfrac{11}{9}$

$\quad\quad\text{RS} = \dfrac{9}{8} - \dfrac{1}{6}\left(-\dfrac{7}{12}\right) = \dfrac{9}{8} + \dfrac{7}{72} = \dfrac{81 + 7}{72} = \dfrac{88}{72} = \dfrac{11}{9}$

(e) $\quad \dfrac{7}{5}(6x - 7) - \dfrac{3}{8}(7x + 15) = 25$

$$56(6x - 7) - 15(7x + 15) = 40(25)$$
$$336x - 392 - 105x - 225 = 1000$$
$$231x - 617 = 1617$$
$$x = 7$$

Check $\text{LS} = \dfrac{7}{5}[6(7) - 7] - \dfrac{3}{8}[7(7) + 15]$

$\quad\quad\quad = \dfrac{7}{5}(35) - \dfrac{3}{8}(64) = 7(7) - 24 = 49 - 24 = 25$

$\quad\quad\text{RS} = 25$

(f) $\quad \dfrac{5}{9}(7 - 6x) - \dfrac{3}{4}(3 - 15x) = \dfrac{1}{12}(3x - 5) - \dfrac{1}{2}$

$$20(7 - 6x) - 27(3 - 15x) = 3(3x - 5) - 18$$

$$140 - 120x - 81 + 405x = 9x - 15 - 18$$

$$285x + 59 = 9x - 33$$

$$276x = -92$$

$$x = -\dfrac{1}{3}$$

Check $\quad$ LS $= \dfrac{5}{9}\left[7 - 6\left(-\dfrac{1}{3}\right)\right] - \dfrac{3}{4}\left[3 - 15\left(-\dfrac{1}{3}\right)\right]$

$$= \dfrac{5}{9}(7 + 2) - \dfrac{3}{4}(3 + 5)$$

$$= 5 - 6 = -1$$

$$\text{RS} = \dfrac{1}{12}\left[3\left(-\dfrac{1}{3}\right) - 5\right] - \dfrac{1}{2}$$

$$= \dfrac{1}{12}(-6) - \dfrac{1}{2}$$

$$= -\dfrac{1}{2} - \dfrac{1}{2} = -1$$

(g) $\quad \dfrac{5}{6}(4x - 3) - \dfrac{2}{5}(3x + 4) = 5x - \dfrac{16}{15}(1 - 3x)$

$$25(4x - 3) - 12(3x + 4) = 150x - 32(1 - 3x)$$

$$100x - 75 - 36x - 48 = 150x - 32 + 96x$$

$$64x - 123 = 246x - 32$$

$$-182x = 91$$

$$x = -\dfrac{1}{2}$$

Check $\quad$ LS $= \dfrac{5}{6}\left[4\left(-\dfrac{1}{2}\right) - 3\right] - \dfrac{2}{5}\left[3\left(-\dfrac{1}{2}\right) + 4\right]$

$$= \dfrac{5}{6}(-2 - 3) - \dfrac{2}{5}\left[-\dfrac{3}{2} + 4\right]$$

$$= \dfrac{5}{6}(-5) - \dfrac{2}{5}\left(\dfrac{5}{2}\right) = -\dfrac{25}{6} - 1 = -\dfrac{31}{6}$$

$$\text{RS} = 5\left(-\dfrac{1}{2}\right) - \dfrac{16}{15}\left[1 - 3\left(-\dfrac{1}{2}\right)\right]$$

$$= -\dfrac{5}{2} - \dfrac{16}{15}\left(1 + \dfrac{3}{2}\right) = -\dfrac{5}{2} - \dfrac{16}{15}\left(\dfrac{5}{2}\right)$$

$$= -\dfrac{5}{2} - \dfrac{8}{3} = \dfrac{-15 - 16}{6} = -\dfrac{31}{6}$$

9. (a)   Let the size of the work force be x.

Number laid off $= \dfrac{1}{6}x$

Number after the layoff $= x - \dfrac{1}{6}x$

$\therefore x - \dfrac{1}{6}x = 690$

$\dfrac{5}{6}x = 690$

$5x = 4140$

$x = 828$

$\therefore$ the number laid off is $\dfrac{1}{6} \times 828 = 138$.

(b)   Let last year's average property value be $\$x$.

Current average value $= \$\left(x + \dfrac{2}{7}x \right)$

$\therefore x + \dfrac{2}{7}x = 81\,450$

$\dfrac{9}{7}x = 81\,450$

$\dfrac{1}{7}x = 9050$

$x = 63\,350$

$\therefore$ last year's average value was $\$63\,350$.

(c)   Let the quoted price be $\$x$.

$\therefore x + \dfrac{1}{20}x = 2457$

$\dfrac{21}{20}x = 2457$

$\dfrac{1}{20}x = 117$

$x = 2340$

$\therefore$ the gratuities $= \dfrac{1}{20}$ of $2340 = \$117$.

(d)   Let the value of the building be $\$x$.

Value of the land $= \$\dfrac{1}{3}x - 2000$

Total value of the property $= \$x + \dfrac{1}{3}x - 2000$

$\therefore x + \dfrac{1}{3}x - 2000 = 184\,000$

$\dfrac{4}{3}x = 186\,000$

$\dfrac{1}{3}x = 46\,500$

$x = 139\,500$

The value assigned to land is $\$(184\,000 - 139\,500) = \$44\,500$.

(e) Let the cost of power be x.

Cost of heat $= \$\left(\dfrac{3}{4}x + 22\right)$

Cost of water $= \$\left(\dfrac{1}{3}x - 11\right)$

Total cost $= x + \dfrac{3}{4}x + 22 + \dfrac{1}{3}x - 11 = 2010 + 10\%$ of 2010.

$$12x + 9x + 4x = 12(2010 + 201 - 11)$$
$$25x = 26\,400$$
$$x = 1056$$

Cost of heat $= \dfrac{3}{4} \times 1056 + 22 = \$814$

Cost of power $= \$1056$

Cost of water $= \dfrac{1}{3} \times 1056 - 11 = \$341$

(f) Let the amount allocated to newspaper advertising be x.
Amount allocated to TV advertising $= \$(3x + 1000)$

Amount allocated to direct selling $= \dfrac{3}{4}[x + 3x + 1000]$

$\therefore x + 3x + 1000 + \dfrac{3}{4}[4x + 1000] = 87\,500$

$$4x + \dfrac{3}{4}[4x + 1000] = 86\,500$$
$$16x + 12x + 3000 = 346\,000$$
$$28x = 343\,000$$
$$x = 12\,250$$

The amount allocated to newspaper advertising is \$12 250; the amount allocated to TV advertising is \$37 750; the amount allocated to direct selling is \$37 500.

(g) Let the number of minutes on Machine B be x.

Time on Machine A $= \dfrac{4}{5}x - 3$ minutes

Time on Machine C $= \dfrac{5}{6}\left(x + \dfrac{4}{5}x - 3\right)$ minutes

Total time $= x + \dfrac{4}{5}x - 3 + \dfrac{5}{6}\left(x + \dfrac{4}{5}x - 3\right)$ minutes

$\therefore x + \dfrac{4}{5}x - 3 + \dfrac{5}{6}\left(x + \dfrac{4}{5}x - 3\right) = 77$

$$30x + 24x - 90 + 25\left(x + \dfrac{4}{5}x - 3\right) = 30(77)$$
$$54x - 90 + 25x + 20x - 75 = 2310$$
$$99x - 165 = 2310$$
$$99x = 2475$$
$$x = 25$$

Time on Machine B is 25 minutes; time on Machine A is $\dfrac{4}{5}(25) - 3 = 17$ minutes;

time on Machine C is $\dfrac{5}{6}(25 + 17) = 35$ minutes.

(h)   Let the number of pairs of superlight poles be x.
Number of pairs of ordinary poles = $72 - x$
Value of superlight poles = $\$30x$
Value of ordinary poles = $\$16(72 - x)$
Total value of all poles = $\$30x + 16(72 - x)$

$$30x + 16(72 - x) = 1530$$
$$30x + 1152 - 16x = 1530$$
$$14x = 378$$
$$x = 27$$

The number of pairs of superlight poles is 27; the number of pairs of ordinary poles is 45.

(i)   Let the number of $2 coins be x.

Number of $1 coins = $\dfrac{3}{5}x + 1$

Number of quarters = $4\left(x + \dfrac{3}{5}x + 1 \right)$

Value of the $2 coins = $\$2x$

Value of the $1 coins = $\$\left[\dfrac{3}{5}x + 1 \right]$

Value of the quarters = $\$\dfrac{1}{4}(4)\left(x + \dfrac{3}{5}x + 1 \right) = x + \dfrac{3}{5}x + 1$

Total value = $2x + \dfrac{3}{5}x + 1 + x + \dfrac{3}{5}x + 1 = 107$

$$10x + 3x + 5 + 5x + 3x + 5 = 535$$
$$21x + 10 = 535$$
$$21x = 525$$
$$x = 25$$

The number of $2 coins is 25; the number of $1 coins is $\left(\dfrac{3}{5} \times 25 + 1 \right) = 16$;

the number of quarters is $4(25 + 16) = 164$.

Self-Test

1. (a) $4 - 3x - 6 - 5x = -2 - 8x$

   (b) $(5x - 4) - (7x + 5) = 5x - 4 - 7x - 5 = -2x - 9$

   (c) $-2(3a - 4) - 5(2a + 3)$
$$= -6a + 8 - 10a - 15$$
$$= -16a - 7$$

   (d) $-6(x - 2)(x + 1)$
$$= -6\left(x^2 - 2x + x - 2 \right)$$
$$= -6\left(x^2 - x - 2 \right)$$
$$= -6x^2 + 6x + 12$$

3. (a) $(-2)^3 = -8$

(b) $\left(-\dfrac{2}{3}\right)^2 = \dfrac{4}{9}$

(c) $(4)^0 = 1$

(d) $(3)^2(3)^5 = (3)^7 = 2187$

(e) $\left(\dfrac{4}{3}\right)^{-2} = \dfrac{1}{\left(\dfrac{4}{3}\right)^2} = \dfrac{1}{\dfrac{16}{9}} = \dfrac{9}{16}$

(f) $\left(-x^3\right)^5 = -x^{15}$

5. (a) $\dfrac{1}{81} = \left(\dfrac{1}{3}\right)^{n-2}$

$\dfrac{1}{3^4} = \left(\dfrac{1}{3}\right)^{n-2}$

$\left(\dfrac{1}{3}\right)^4 = \left(\dfrac{1}{3}\right)^{n-2}$

Since the bases are common
$4 = n - 2$
$n = 6$

(b) $\dfrac{5}{2} = 40\left(\dfrac{1}{2}\right)^{n-1}$

$\dfrac{1}{2} = 8\left(\dfrac{1}{2}\right)^{n-1}$

$\dfrac{1}{16} = \left(\dfrac{1}{2}\right)^{n-1}$

$\left(\dfrac{1}{2}\right)^4 = \left(\dfrac{1}{2}\right)^{n-1}$

$4 = n - 1$
$n = 5$

7. (a) $I = Prt$

$$P = \frac{I}{rt}$$

(b) $$S = \frac{P}{1 - dt}$$

$$\frac{S}{P} = \frac{1}{1 - dt}$$

$$\frac{P}{S} = 1 - dt$$

$$dt = 1 - \frac{P}{S}$$

$$d = \frac{1 - \dfrac{P}{S}}{t}$$

$$d = \frac{\dfrac{S - P}{S}}{t}$$

$$d = \frac{S - P}{St}$$

3 Ratio, proportion and percent

Exercise 3.1

A. 1. (a) $12:32 = 3:8$

(b) $84:56 = 3:2$

(c) $15:24:39 = 5:8:13$

(d) $21:42:91 = 3:6:13$

3. (a) $\dfrac{1.25}{4} = \dfrac{125}{400} = \dfrac{5}{16}$

(b) $\dfrac{2.4}{8.4} = \dfrac{24}{84} = \dfrac{2}{7}$

(c) $0.6:2.1:3.3$
$= 6:21:33$
$= 2:7:11$

(d) $5.75:3.50:1.25$
$= 575:350:125$
$= 23:14:5$

(e) $\dfrac{1}{2}:\dfrac{2}{5} = 5:4$

(f) $\dfrac{5}{3}:\dfrac{7}{5} = 25:21$

(g) $\dfrac{3}{8}:\dfrac{2}{3}:\dfrac{3}{4}$
$= 9:16:18$

(h) $\dfrac{2}{5}:\dfrac{4}{7}:\dfrac{5}{14}$
$= 28:40:25$

(i) $2\dfrac{1}{5}:4\dfrac{1}{8}$

$= \dfrac{11}{5}:\dfrac{33}{8}$

$= 88:165$

$= 8:15$

(j) $5\dfrac{1}{4}:5\dfrac{5}{6}$

$= \dfrac{21}{4}:\dfrac{35}{6}$

$= 63:70$

$= 9:10$

B. 1. $\dfrac{\text{Food Cost}}{\text{Beverage Cost}} = \dfrac{40\%}{35\%} = \dfrac{8}{7}$

3. Supervisors : office employees : production workers
$=\quad 6 \quad : \qquad 9 \qquad : \qquad\qquad 36$
$=\ 2:3:12$

C. 1. A:B:C: = 9:2:1
Total number of shares is 12.
Value of each share = 3060 ÷ 12 = 255
A receives 9 × 255 = $2295
B receives 2 × 255 = $ 510
C receives 1 × 255 = $ 255

3. Ratio $= \dfrac{5}{8}:\dfrac{1}{3}:\dfrac{1}{6} = 15:8:4$
Total number of parts = 27
Value of each part = 9450 ÷ 27 = $350
Distribution:
   Manufacturing: 15 × 350 = $5250
   Selling: 8 × 350 = $2800
   Administration: 4 × 350 = $1400

Exercise 3.2

1. $3:n = 15:20$
    $15n = 20 \times 3$
    $n = \dfrac{20 \times 3}{15}$
    $n = 4$

3. $3:8 = 21:x$
    $3x = 21 \times 8$
    $x = \dfrac{21 \times 8}{3}$
    $x = 56$

5. $1.32:1.11 = 8.8:k$
    $1.32k = 1.11 \times 8.8$
    $k = \dfrac{1.11 \times 8.8}{1.32}$
    $k = 7.4$

7. $m:3.4 = 2.04:2.89$
    $2.89m = 2.04 \times 3.4$
    $m = \dfrac{2.04 \times 3.4}{2.89}$
    $m = 2.4$

9. $t:\dfrac{3}{4} = \dfrac{7}{8}:\dfrac{15}{16}$
    $\dfrac{15}{16}t = \dfrac{3}{4} \times \dfrac{7}{8}$
    $t = \dfrac{3}{4} \times \dfrac{7}{8} \times \dfrac{16}{15}$
    $t = \dfrac{7}{10}$

11. $\dfrac{9}{8}:\dfrac{3}{5} = t:\dfrac{8}{15}$
    $\dfrac{3}{5}t = \dfrac{8}{15} \times \dfrac{9}{8}$
    $t = \dfrac{8}{15} \times \dfrac{9}{8} \times \dfrac{5}{3}$
    $t = 1$

B. 1. Let the number of months to earn $8.75 per share be x.
    $$\therefore \$1.25:3 = 8.75:x$$
    $$1.25x = 3(8.75)$$
    $$x = 21$$

3. Let the distance travelled on 75 litres be x km.
    $$\therefore 9 \text{ l}:72 \text{ km} = 75 \text{ l}:x \text{ km}$$
    $$9x = 72(75)$$
    $$x = 600$$

5. (a) Let the total value before selling be x.
    $$5:6 = 3000:x$$
    $$5x = 18\ 000$$
    $$x = 3600$$

    (b) Let the value of the partnership be y.
    $$2:5 = 3600:y$$
    $$2y = 18\ 000$$
    $$y = 9000$$

7. Let last year's profit be x.
    $$\frac{4}{9} = \frac{\$12\ 800 \text{ dividend}}{\$x \text{ profit}}$$
    $$4x = 9(12\ 800)$$
    $$x = 28\ 800$$
    Let revenue be y.
    $$\frac{2}{7} = \frac{28\ 800}{y}$$
    $$2y = 7(28\ 800)$$
    $$y = 100\ 800$$
    Last year's revenue was $100 800.

Exercise 3.3

A. 1. $0.40 \times 90 = 36$

3. $2.50 \times 120 = 300$

5. $0.03 \times 600 = 18$

7. $0.005 \times 1200 = 6$

9.   0.0002 × 2500 = 0.5

11.   0.0025 × 800 = 2

13.   0.00075 × 10 000 = 7.5

15.   0.025 × 700 = 17.5

B.   1. $\frac{1}{3}$ × 48 = \$16

3. $\frac{13}{8}$ × 1200 = 13 × 150 = \$1950

5. $\frac{3}{8}$ × 24 = 3 × 3 = \$9

7. $\frac{5}{4}$ × 160 = 5 × 40 = \$200

9. $\frac{5}{6}$ × 720 = 5 × 120 = \$600

11. $\frac{7}{6}$ × 42 = 7 × 7 = \$49

13. $\frac{3}{4}$ × 180 = 3 × 45 = \$135

15. $\frac{4}{3}$ × 45 = 4 × 15 = \$60

C.   1.   1% → 31.20
     $\frac{1}{2}$% → \$15.60

3. $1\% \rightarrow 4.32$

$\dfrac{1}{8}\% \rightarrow 0.54$

$\dfrac{3}{8}\% \rightarrow \$1.62$

5. $1\% \rightarrow 36.30$

$\dfrac{1}{3}\% \rightarrow 12.10$

$1\dfrac{1}{3}\% \rightarrow \$48.40$

7. $1\% \rightarrow 9.44;\ \dfrac{1}{8}\% = 1.18$

$2\% \rightarrow 18.88;\ \dfrac{3}{8}\% = 3.54$

$2\dfrac{3}{8}\% \rightarrow \$22.42$

D. 1. $x + 0.4x = 28$
$$1.4x = 28$$
$$x = 20$$

3. $x - 0.05x = 418$
$$0.95x = 418$$
$$x = 440$$

5. $x + \dfrac{1}{6}x = 42$
$$\dfrac{7x}{6} = 42$$
$$x = 36$$

7. $x + 1.5x = 75$
$$2.5x = 75$$
$$x = 30$$

E. 1. $R = \dfrac{36}{60} = 0.60 = 60\%$

3. $R = \dfrac{920}{800} = 1.15 = 115\%$

5. $R = \dfrac{6}{120} = \dfrac{1}{20} = 5\%$

7. $R = \dfrac{132}{22} = 6 = 600\%$

9. $R = \dfrac{150}{90} = \dfrac{5}{3} = 166\dfrac{2}{3}\%$

F. 1. $60 = 30\%$ of x
    $60 = 0.3x$
    $x = 200$

3. $x = 0.1\%$ of 3600
    $= 0.001 \times 3600$
    $= 3.60$

5. $x = \dfrac{1}{2}\%$ of 612
    $1\% \rightarrow 6.12$
    $\dfrac{1}{2}\% \rightarrow 3.06$

7. $80 = 40\%$ of x
    $80 = 0.4x$
    $x = 200$

9. $x = \dfrac{1}{8}\%$ of 880
    $1\% = 8.80$
    $\dfrac{1}{8}\% = 1.10$

11.　　$600 = 250\%$ of x
$$2.5x = 600$$
$$x = 240$$

13.　　$R = \dfrac{350}{70} = 5 = 500\%$

15.　　350% of $x = 1050$
$$3.5x = 1050$$
$$x = 300$$

G.　1.　Let the reduction be x.
$$x = 40\% \text{ of } 70$$
$$x = 28$$

3.　Let waste be x.
$$x = 6\% \text{ of } 25\,000$$
$$= 0.06 \times 25\,000$$
$$= 1500$$

5.　Let the budgeted sales be x.
$$90\% \text{ of } x = 40\,500$$
$$0.9x = 40\,500$$
$$x = 45\,000$$

7.　Let population 5 years ago be x.
$$120\% \text{ of } x = 54\,000$$
$$1.2x = 54\,000$$
$$x = 45\,000$$

Exercise 3.4

A.　1.　$x = 120 + 40\%$ of 120
$$= 120 + 0.40 \times 120$$
$$= 168$$

3.　$x = 1200 - 5\%$ of 1200
$$= 1200 - 0.05 \times 1200$$
$$= 1140$$

5. $x = 48 + 83\frac{1}{3}\% \text{ of } 48$

    $= 48 + \frac{5}{6} \times 48$

    $= 88$

B. 1. Increase = 15

    $R = \frac{15}{30} = \frac{1}{2} = 50\%$

3. Increase = 160

    $R = \frac{160}{80} = 2 = 200\%$

5. Decrease = 6

    $R = \frac{6}{300} = 0.02 = 2\%$

C. 1. $24 = x - 25\% \text{ of } x$

    $24 = x - \frac{1}{4}x$

    $\frac{3}{4}x = 24$

    $\frac{1}{4}x = 8$

    $x = 32$

3. $x + 150\% \text{ of } x = 325$

    $x + 1.5x = 325$

    $2.5x = 325$

    $x = 130$

5. $x - 5\% \text{ of } x = 4.18$

    $0.95x = 4.18$

    $x = 4.40$

Exercise 3.5

A. 1. Let the number of absentees be x.

$$x = 2\frac{1}{4}\% \text{ of } 1200$$

$$1\% \rightarrow 12$$
$$2\% \rightarrow 24$$
$$\frac{1}{4}\% \rightarrow 3$$
$$2\frac{1}{4}\% \rightarrow 27$$

The number absent is 27.

3. Increase $= \dfrac{35}{280} = \dfrac{5}{40} = \dfrac{1}{8} = 12.5\%$

5. Let the weekly sales be $\$x$.

$$16\frac{2}{3}\% \text{ of } x = 720$$

$$\frac{1}{6}x = 720$$

$$x = 4320$$

Weekly sales must be $4320.

7. (a) Let the policy value be $\$x$.

$$\frac{3}{8}\% \text{ of } x = 675$$

$$\frac{1}{8}\% \text{ of } x = 225$$

$$1\% \text{ of } x = 1800$$

Face value is $180 000.

(b) Let appraised value be $\$y$.

$$80\% \text{ of } y = 180\,000$$
$$0.8y = 180\,000$$
$$y = 225\,000$$

Appraised value is $225 000.

B. 1. Let the selling price be $\$x$.

$$7.92 + 83\frac{1}{3}\% \text{ of } 7.92 = x$$

$$7.92 + \frac{5}{6} \times 7.92 = x$$

$$7.92 + 6.60 = x$$

$$x = 14.52$$

The article sold for $14.52.

3. Let the reduced price be x.

$$x = 225 - 33\frac{1}{3}\% \text{ of } 225$$

$$= 225 - \frac{1}{3} \times 225$$

$$= 225 - 75$$

$$= 150$$

The reduced price was $150.

5. Let the cost before blended sales tax be x.

$$15\% \text{ of } x = 11.10$$

$$0.15x = 11.10$$

$$x = 74.00$$

The amount paid $= 74.00 + 11.10 = \$85.10$.

7. Let the face value be x.

$$12\frac{1}{2}\% \text{ of } x = 625$$

$$\frac{1}{8}x = 625$$

$$x = 5000$$

The face value is $5000.

9. Decrease $= 6540 - 1090 = 5450$

Rate $= \dfrac{5450}{6540} = 0.83\frac{1}{3} = 83\frac{1}{3}\%$

Decrease in profit was $83\frac{1}{3}\%$.

11. Gain in value $= 178\,500 - 42\,000 = 136\,500$

Rate $= \dfrac{136\,500}{42\,000} = 3.25 = 325\%$

13. Let the marked price be x.

$$x - 33\frac{1}{3}\% \text{ of } x = 64.46$$

$$x - \frac{1}{3}x = 64.46$$

$$\frac{2}{3}x = 64.46$$

$$\frac{1}{3}x = 32.23$$

$$x = 96.69$$

The marked price was $96.69.

15. Let the invoice amount be x.
$$x - 5\% \text{ of } x = 646$$
$$0.95x = 646$$
$$x = 680$$
The invoice amount was $680.

17. Let the second quarter working capital be x.
$$x + 75\% \text{ of } x = 78\,400$$
$$\frac{7}{4}x = 78\,400$$
$$\frac{1}{4}x = 11\,200$$
$$x = 44\,800$$
Working capital at the end of the second quarter was $44 800.

19. Let the compensation before vacation pay be x.
$$x + 4\% \text{ of } x = 23\,400$$
$$1.04x = 23\,400$$
$$x = 22\,500$$
Vacation Pay $= 4\% \text{ of } 22\,500 = \$900$

Exercise 3.6

A. 1. Let the number of $U.S. be x.
$$\frac{\$1 \text{ CDN}}{\$0.6852 \text{ U.S.}} = \frac{\$750.00 \text{ CDN}}{\$x \text{ U.S.}}$$
$$\frac{1}{0.6852} = \frac{750.00}{x}$$
$$x = 750.00(0.6852)$$
$$x = 513.90$$
$750.00 CDN will buy $513.90 U.S.

3. Let the number of $CDN be x.
$$\frac{\$1 \text{ CDN}}{\$0.67 \text{ U.S.}} = \frac{\$x \text{ CDN}}{\$149.00 \text{ U.S.}}$$
$$\frac{1}{0.67} = \frac{x}{149.00}$$
$$x = \frac{149.00}{0.67}$$
$$x = 222.39$$
The flight costs $222.39 CDN.

B. 1. $1 U.S. = $1.5134 CDN
        $350.00 U.S. convert to 350.00(1.5134) = $529.69

    3. $1 U.S. = 1.4829 Swiss francs
        $175.00 U.S. convert to $175.00(1.4829) = 259.51 Swiss francs

    5. 1 Euro = 1.9558 German marks
        550 Euros convert to 550.00(1.9558) = 1075.69 German marks

Exercise 3.7

A. 1. Simple price index for bread $= \dfrac{1.44}{0.92}(100) = 156.52$

        Simple price index for tires $= \dfrac{94.00}{85.00}(100) = 110.59$

        Simple price index for computers $= \dfrac{1450.00}{2100.00}(100) = 69.05$

        Interpretation:
            The price of bread increased 56.52% from 1991 to 2001.
            The price of tires increased 10.59% from 1991 to 2001.
            The price of computers decreased 30.95% from 1991 to 2001.

B. 1. (a) Relative to 1992,
            Purchasing power of the dollar in 1996 $= \dfrac{1}{106.8}(100) = \$0.94$

            Purchasing power of the dollar in 1999 $= \dfrac{1}{109.1}(100) = \$0.92$

        (b) Relative to 1996,
            Purchasing power of the dollar in 1999 $= \dfrac{106.8}{109.1}(100) = \$0.98$

    3.
        $$\frac{\text{Earnings 1991}}{\text{CPI 1991}} = \frac{\text{Earnings 1999}}{\text{CPI 1999}}$$
        $$\frac{32\,500}{98.5} = \frac{x}{109.1}$$
        $$x = \frac{32\,500(109.1)}{98.5} = 35\,997.46$$

        Tamara's 1999 earnings will have to be $35 997.46.

Exercise 3.8

A. 1. Federal tax on first $29 590 $5 030.00
        Federal tax on excess
            26% of $(49 450 − 29 590) = 0.26(19 860)   5 163.60
                    Federal tax reported $10 193.60

    3. Taxable income = 32 920 + 7700 = $40 620
        Federal tax on first $29 590 $5 030.00
        Federal tax on excess
            26% of $(40 620 − 29 590) = 0.26(11 030)   2 867.80
                    Federal tax reported $7 897.80

Exercise 3.9

A. 1. (a) Simple rate of return for EBF, February 29, 2000
            $= \dfrac{16.54 - 16.08}{16.08} = \dfrac{0.46}{16.08} = 0.0286070 = 2.86\%$

        (b) Simple rate of return for EBF, March 10, 2000
            $= \dfrac{16.95 - 16.08}{16.08} = \dfrac{0.87}{16.08} = 0.0541045 = 5.41\%$

    3. Percent change from March 21 to March 22, 2000
        $= \dfrac{55.90 - 58.90}{58.90} = \dfrac{-3.00}{58.90} = -0.0509338 = -5.09\%$

Review Exercise

1. (a) 25 dimes:3 dollars = 250:300 = 5:6

    (b) 5 hours:50 minutes = 300:50 = 6:1

    (c) $6.75:30 litres = 675:3000 = 135:600 = 9:40

    (d) $21:3.5 hours = 210:35 = 30:5 = 6:1

    (e) 1440 words:120 lines:6 pages = 240:20:1

    (f) 90 kg:24 ha:18 weeks = 15:4:3

3. (a)   150% of 140

$= 1.5 \times 140 = 210$

(b)   3% of 240

$= 0.03 \times 240 = 7.2$

(c) $9\frac{3}{4}$% of 2000

$= 0.0975 \times 2000 = 195$

(d)   0.9% of 400

$= 0.009 \times 400 = 3.6$

5. (a)   1% of \$2664 $=$ \$26.64

$\frac{1}{4}\% \rightarrow$ \$6.66

(b)   1% of \$1328 $=$ \$13.28

$\frac{1}{8}\% \rightarrow 1.66$

$\frac{5}{8}\% \rightarrow$ \$8.30

(c)   1% of \$5400 $=$ \$54.00

$\frac{1}{3}\% \rightarrow 18.00$

$\frac{2}{3}\% \rightarrow 36.00$

$1\frac{2}{3}\% \rightarrow$ \$90.00

(d)   1% of \$1260 $=$ \$12.60

$2\% \rightarrow 25.20$

$\frac{1}{5}\% \rightarrow 2.52$

$2\frac{1}{5}\% \rightarrow$ \$27.72

7. (a) $8 + 125\%$ of $8 = x$
              $8 + 10 = x$
                   $x = \$18$

    (b) $x = 2000 - 2\frac{1}{4}\%$ of 2000
            $= 2000 - 45$
            $= \$1955$

    (c)   Decrease $= 120 - 100 = 20$
        Rate of decrease $= \dfrac{20}{120} = \dfrac{1}{6} = 16\frac{2}{3}\%$.

    (d)   Increase $= 975 - 150 = 825$
        Rate of increase $= \dfrac{825}{150} = 5.50 = 550\%$.

    (e) $98 = x + 75\%$ of x
       $1.75x = 98$
          $x = 56$
       The amount is $\$56$.

    (f) $x - 15\%$ of $x = 289$
           $0.85x = 289$
              $x = 340$
       The price was $\$340$.

    (g) $x + 250\%$ of $x = 490$
           $x + 2.5x = 490$
             $3.5x = 490$
                $x = 140$
       The sum of money is $\$140$.

9. Total number of parts $= 80 + 140 + 160 = 380$
Rental allocation per part $= 11\,400/380 = 30$
Allocation:
       Department A: $80 \times 30 = \$2400$
       Department B: $140 \times 30 = \$4200$
       Department C: $160 \times 30 = \$4800$

11. Ratio $= \dfrac{1}{2}:\dfrac{1}{3}:\dfrac{2}{5} = \dfrac{15}{30}:\dfrac{10}{30}:\dfrac{12}{30} = 15:10:12$

$15x + 10x + 12x = 185\,000$

$x = 5\,000$

Allocation of fire loss:

Company 1: $15 \times 5000 = \$75\,000$

Company 2: $10 \times 5000 = \$50\,000$

Company 3: $12 \times 5000 = \$60\,000$

13. Let the variable cost for sales of \$350 000 be \$$x$.

$$\frac{\$130\,000 \text{ variable cost}}{\$250\,000 \text{ sales}} = \frac{\$x \text{ variable cost}}{\$350\,000 \text{ sales}}$$

$$\frac{130\,000}{250\,000} = \frac{x}{350\,000}$$

$$25x = 13(350\,000)$$

$$x = \frac{13(350\,000)}{25}$$

$$x = 182\,000$$

The variable cost for sales of \$350 000 is \$182 000.

15. $\dfrac{\text{Faculty}}{\text{Support}} = \dfrac{5}{4} = \dfrac{x}{192}$

$192(5) = 4x$

$x = 240$

The number of faculty members is 240.

$\dfrac{4}{9}$ of total $= 240$

$\dfrac{4}{9}T = 240$

$4T = 240 \times 9$

$T = \dfrac{240 \times 9}{4}$

$T = 540$

Employment is 540.

17. Bonds $= 37\dfrac{1}{2}\% \text{ of } 150\,000$

$= \dfrac{3}{8} \times 150\,000 \qquad = \$\ 56\,250$

Common $= 56\dfrac{1}{4}\% \text{ of } 150\,000$

$= 0.5625 \times 150\,000 = \$\ 84\,375$

Preferred $= (100\% - 37.5\% - 56.25\%)$

$= 6.25\% \text{ of } 150\,000$

$= 0.0625 \times 150\,000 = \underline{\$\ \ \ 9\,375}$

Total $= \$150\,000$

19. (a) Appraised value

$$= 120\,000 + 233\tfrac{1}{3}\% \text{ of } 120\,000$$

$$= 120\,000 + \frac{7}{3} \times 120\,000$$

$$= 120\,000 + 280\,000$$

$$= \$400\,000$$

(b) Gain $= \$280\,000$

21. (a) Not passed $= \dfrac{180}{2400} = \dfrac{3}{40} = 7.5\%$

(b) Scrapped as a percent of production not passed

$$= \frac{30}{180} = \frac{1}{6} = 16\tfrac{2}{3}\%$$

23. (a) Increase in pay
$= 16.80 - 6.30 = \$10.50$
Percent change

$$= \frac{10.50}{6.30} = 1.\dot{6} = 166\tfrac{2}{3}\%$$

(b) Current pay as a percent of old pay

$$= \frac{16.80}{6.30} = 2.\dot{6} = 266\tfrac{2}{3}\%$$

25. List price $= 240\%$ of cost
$396 = 2.40C$
$C = 165$
Cost is \$165.

27. 77.5% of list price $=$ sale price
$0.775L = 15\,500$
$L = 20\,000$

List $= C + 33\tfrac{1}{3}\%$ of C

$$20\,000 = \frac{4}{3}C$$

$C = 15\,000$
Cost was \$15 000.

29. (a) Let the asking price be x.

$$\therefore 91\frac{2}{3}\% \text{ of } x = 330\,000$$

$$\frac{11}{12}x = 330\,000$$

$$\frac{1}{12}x = 30\,000$$

$$x = 360\,000$$

Let the cost be y.

$$y + 350\% \text{ of } y = 360\,000$$

$$4.5y = 360\,000$$

$$y = 80\,000$$

The original cost was $80\,000.

(b) Gain $= 330\,000 - 80\,000 = \$250\,000$

(c) Percent gain $= \dfrac{250\,000}{80\,000} = 3.125 = 312.5\%$

31. Let the value of the coupon be x U.S.

$$\frac{\$x \text{ U.S.}}{\$216.00 \text{ CDN}} = \frac{\$1 \text{ U.S.}}{\$1.49 \text{ CDN}}$$

$$\frac{x}{216.00} = \frac{1}{1.49}$$

$$x = \frac{216.00}{1.49} = 144.97$$

The coupon has a value of $144.97 U.S.

33.

Federal tax on first $59 180	$12 724.00
Federal tax on excess	
29% of $(83 450 − 59 180) = 0.29(24 270)	7 038.30
Federal tax	$19 762.30

Self-Test

1. (a) 125% of 280 $= \dfrac{5}{4} \times 280 = 5 \times 70 = \$350$

(b) $\dfrac{3}{8}\%$ of $20 280: $1\% \rightarrow \$202.80$

$$\frac{1}{8}\% \rightarrow \$25.35$$

$$\frac{3}{8}\% \rightarrow \$76.05$$

(c) $83\frac{1}{3}\%$ of $\$174 = \frac{5}{6} \times 174 = 5 \times 29 = \$145.00$

(d) $1\frac{1}{4}\%$ of $\$1056$:
$$1\% \rightarrow \$10.56$$
$$\frac{1}{4}\% \rightarrow \$\ 2.64$$
$$\overline{1\frac{1}{4}\% \rightarrow \$13.20}$$

3. Total in sample $= 24 + 36 + 20 = 80$
Brand Y preference $= \dfrac{36}{80} = \dfrac{9}{20} = 0.45 = 45\%$

5. $\dfrac{\text{Beverage sales}}{\text{Food sales}} = \dfrac{\$9.60}{\$12.00} = \dfrac{96}{120} = \dfrac{4}{5}$

Let the budget for beverage sales be $\$x$.

$$\frac{x}{12\ 500} = \frac{4}{5}$$
$$x = \frac{4}{5} \times 12\ 500$$
$$x = 10\ 000$$

The monthly budget for beverage sales is $\$10\ 000$.

7. $\dfrac{1}{2}:\dfrac{1}{3}:\dfrac{1}{5}:\dfrac{1}{6} = \dfrac{15}{30}:\dfrac{10}{30}:\dfrac{6}{30}:\dfrac{5}{30} = 15:10:6:5$

Total number of shares $= 15 + 10 + 6 + 5 = 36$
Value of each share $= 40\ 500 \div 36 = \$1125$
First bonus: $\quad 15 \times 1125 = \$16\ 875$
Second bonus: $\quad 10 \times 1125 = \$11\ 250$
Third bonus: $\quad\ \ 6 \times 1125 = \$\ 6\ 750$
Fourth bonus: $\quad\ 5 \times 1125 = \$\ 5\ 625$

9. Let the price of the article before taxes be $\$x$; then the amount of blended sales tax is 15% of $\$x$; the sales value is $x + 15\%$ of x.
$$x + 15\% \text{ of } x = 287.50$$
$$x + 0.15x = 287.50$$
$$1.15x = 287.50$$
$$x = \frac{287.50}{1.15}$$
$$x = 250$$
Blended sales tax $= 15\%$ of $250 = \$37.50$

11. Let the index ten years ago be x; then the increase is 100% of x.

$$x + 100\% \text{ of } x = 360$$
$$x + x = 360$$
$$2x = 360$$
$$x = 180$$

The index 10 years ago was 180.

13. (a) Let the number of German marks be x.

$$\frac{x \text{ German marks}}{\$1 \text{ CDN}} = \frac{1 \text{ German mark}}{\$0.821 \text{ CDN}}$$
$$\frac{x}{1} = \frac{1}{0.821}$$
$$x = 1.2180$$

$1 CDN costs 1.2180 German marks.

(b) To buy $500 CDN costs $500(1.2180) = 609$ German marks.

15. Purchasing power of dollar relative to 1992

$$= \frac{1}{113.6}(100) = \$0.88$$

17. (a) The percent change from January 31 to March 13, 2000

$$\frac{4.55 - 4.60}{4.60} = \frac{-0.05}{4.60} = -0.0108696 = -1.09\%$$

(b) The simple rate of return from January 31 to March 13, 2000

$$\frac{4.55 - 4.60}{4.60} = \frac{-0.05}{4.60} = -0.0108696 = -1.09\%$$

The net asset value dropped 1.09% from January 31 to March 13, 2000.

The answers to questions (a) and (b) are the same. By definition the simple rate of return (not annualized) is the percent change over a period of time.

4 Linear systems

Exercise 4.1

A. 1. $x + y = -9$ ①
   $x - y = -7$ ②

    ① + ② → $2x = -16$
                  $x = -8$

    In ① $-8 + y = -9$
                    $y = -1$
              $(x,y) = (-8,-1)$

    Check:
    In ①   LS $= -8 - 1 = -9 =$ RS
    In ②   LS $= -8 - (-1) = -7 =$ RS

3. $5x + 2y = 74$ ①
   $7x - 2y = 46$ ②

    ① + ② → $12x = 120$
                   $x = 10$

    In ① $5(10) + 2y = 74$
                 $2y = 24$
                  $y = 12$
             $(x,y) = (10,12)$

    Check:
    In ①   LS $= 5(10) + 2(12) = 50 + 24 = 74 =$ RS
    In ②   LS $= 7(10) - 2(12) = 70 - 24 = 46 =$ RS

5. $y = 3x + 12$ ①
   $x = -y$ ②

    Rearrange
                $3x - y = -12$ ③
                $x + y = 0$ ④

    ③ + ④ → $4x = -12$
                   $x = -3$

    In ② $-3 = -y$
                 $y = 3$
             $(x,y) = (-3,3)$

    Check:
    In ①   LS $= -3$
            RS $= 3(-3) + 12 = -9 + 12 = 3$
    In ②   LS $= -3$
            RS $= -3$

CHAPTER 4

B. 1. $4x + y = -13$ ①
      $x - 5y = -19$ ②

To eliminate y

    ① $\times 5 \rightarrow$ $20x + 5y = -65$ ③
                              $x - 5y = -19$ ②
    ③ + ② $\rightarrow$ $21x = -84$
                                 $x = -4$
    In ① $4(-4) + y = -13$
                     $-16 + y = -13$
                           $y = 3$
                     $(x,y) = (-4,3)$

Check:
In ①   LS $= 4(-4) + 3 = -16 + 3 = -13 =$ RS
In ②   LS $= -4 - 5(3) = -4 - 15 = -19 =$ RS

3. $7x - 5y = -22$ ①
   $4x + 3y = 5$ ②

To eliminate y

   ① $\times 3 \rightarrow$ $21x - 15y = -66$ ③
   ② $\times 5 \rightarrow$ $20x + 15y = 25$ ④
   ③ + ④ $\rightarrow$ $41x = -41$
                           $x = -1$
   In ② $-4 + 3y = 5$
                   $3y = 9$
                   $y = 3$
              $(x,y) = (-1,3)$

Check:
In ①   LS $= 7(-1) - 5(3) = -7 - 15 = -22 =$ RS
In ②   LS $= 4(-1) + 3(3) = -4 + 9 = 5 =$ RS

5. $12y = 5x + 16$ ①
   $6x + 10y - 54 = 0$ ②

Rearrange

           $-5x + 12y = 16$ ③
           $6x + 10y = 54$ ④

To eliminate x

   ③ $\times 6 \rightarrow$ $-30x + 72y = 96$ ⑤
   ④ $\times 5 \rightarrow$ $30x + 50y = 270$ ⑥
   ⑤ + ⑥ $\rightarrow$ $122y = 366$
                         $y = 3$
   In ① $12(3) = 5x + 16$
              $36 - 16 = 5x$
                   $5x = 20$
                   $x = 4$
             $(x,y) = (4,3)$

Check:
In ①   LS $= 12(3) = 36$
        RS $= 5(4) + 16 = 20 + 16 = 36$
In ②   LS $= 6(4) + 10(3) - 54 = 24 + 30 - 54 = 0$
        RS $= 0$

C. 1. $\quad 0.4x + 1.5y = 16.8$ ①
   $\quad\quad 1.1x - 0.9y = 6.0$ ②

   To eliminate decimals
   ① $\times 10 \rightarrow \quad 4x + 15y = 168$ ③
   ② $\times 10 \rightarrow \quad 11x - 9y = 60$ ④
   To eliminate y
   ③ $\times 3 \rightarrow \quad 12x + 45y = 504$
   ④ $\times 5 \rightarrow \quad 55x - 45y = 300$
   Add: $\quad\quad\quad\quad\quad 67x = 804$
   $\quad\quad\quad\quad\quad\quad\quad\quad x = 12$
   In ③ $\quad\quad 4(12) + 15y = 168$
   $\quad\quad\quad\quad\quad 48 + 15y = 168$
   $\quad\quad\quad\quad\quad\quad\quad 15y = 120$
   $\quad\quad\quad\quad\quad\quad\quad\quad y = 8$
   $\quad\quad\quad\quad\quad (x,y) = (12,8)$

3. $\quad 2.4x + 1.6y = 7.60$ ①
   $\quad\quad 3.8x + 0.6y = 7.20$ ②

   To eliminate decimals
   ① $\times 5 \rightarrow \quad 12x + 8y = 38$ ③
   ② $\times 5 \rightarrow \quad 19x + 3y = 36$ ④
   To eliminate y
   ③ $\times 3 \rightarrow \quad 36x + 24y = 114$
   ④ $\times (-8) \rightarrow -152x - 24y = -288$
   Add: $\quad\quad\quad\quad -116x = -174$
   $\quad\quad\quad\quad\quad\quad\quad\quad x = 1.50$
   In ③ $\quad\quad 12(1.5) + 8y = 38$
   $\quad\quad\quad\quad\quad 18 + 8y = 38$
   $\quad\quad\quad\quad\quad\quad\quad 8y = 20$
   $\quad\quad\quad\quad\quad\quad\quad\quad y = 2.5$
   $\quad\quad\quad\quad\quad (x,y) = (1.5,2.5)$

5. $\quad \dfrac{3x}{4} - \dfrac{2y}{3} = \dfrac{-13}{6}$ ①

   $\quad \dfrac{4x}{5} + \dfrac{3y}{4} = \dfrac{123}{10}$ ②

   To eliminate fractions
   ① $\times 12 \rightarrow \quad 9x - 8y = -26$ ③
   ② $\times 20 \rightarrow \quad 16x + 15y = 246$ ④
   To eliminate y
   ③ $\times 15 \rightarrow \quad 135x - 120y = -390$
   ④ $\times 8 \rightarrow \quad 128x + 120y = 1968$
   Add: $\quad\quad\quad\quad\quad 263x = 1578$
   $\quad\quad\quad\quad\quad\quad\quad\quad x = 6$
   In ③ $\quad\quad 9(6) - 8y = -26$
   $\quad\quad\quad\quad\quad 54 - 8y = -26$
   $\quad\quad\quad\quad\quad\quad -8y = -80$
   $\quad\quad\quad\quad\quad\quad\quad\quad y = 10$
   $\quad\quad\quad\quad\quad (x,y) = (6,10)$

7. $\dfrac{x}{3} + \dfrac{2y}{5} = \dfrac{7}{15}$ ①

$\dfrac{3x}{2} - \dfrac{7y}{3} = -1$ ②

To eliminate fractions

① × 15 → $5x + 6y = 7$ ③

② × 6 → $9x - 14y = -6$ ④

To eliminate y

③ × 7 → $35x + 42y = 49$

④ × 3 → $27x - 42y = -18$

Add: $62x = 31$

$x = \dfrac{1}{2}$

In ③ $5\left(\dfrac{1}{2}\right) + 6y = 7$

$\dfrac{5}{2} + 6y = 7$

$5 + 12y = 14$

$12y = 9$

$y = \dfrac{3}{4}$

$(x,y) = \left(\dfrac{1}{2}, \dfrac{3}{4}\right)$

Exercise 4.2

A. 1.   A(–4,–3)

B(0,–4)

C(3,–4)

D(2,0)

E(4,3)

F(0,3)

G(–4,4)

H(–5,0)

3. (a)

x	–5	–4	–3	–2	–1	0	1	2	3
y	–3	–2	–1	0	1	2	3	4	5

(b)

x	3	2	1	0	–1	–2
y	5	3	1	–1	–3	–5

(c)

x	3	2	1	0	–1	–2	–3
y	6	4	2	0	–2	–4	–6

(d)

x	–5	–4	–3	–2	–1	0	1	2	3	4	5
y	5	4	3	2	1	0	–1	–2	–3	–4	–5

B. 1.

x	0	3	2
y	–3	0	–1

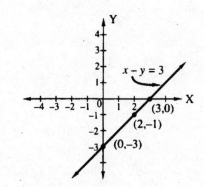

3.

x	0	–2	2
y	0	2	–2

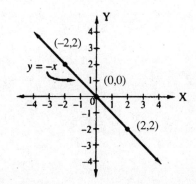

5.

x	0	4	–4
y	–3	0	–6

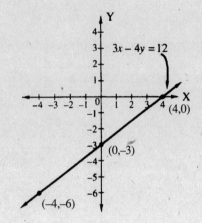

7.

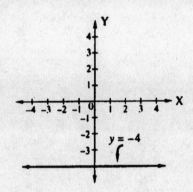

9. For $y = 2x - 3$
    slope, $m = 2$
    y-intercept, $b = -3$

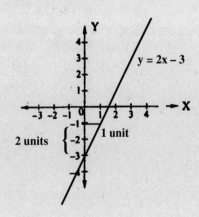

C. 1. For $y = 3x + 20$

x	0	20	40
y	20	80	140

or   m = 3
      b = 20

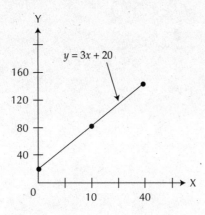

3. For $3x + 4y = 1200$

x	0	200	400
y	300	150	0

or   m $= -\dfrac{3}{4}$
      b = 300

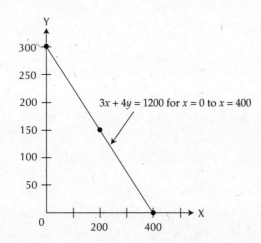

Exercise 4.3

A. 1.

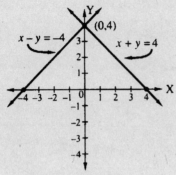

(0,4) is the solution.

3. $x = 2y - 1$

x	3	–1	–5
y	2	0	–2

$y = 4 - 3x$

x	0	2	1
y	4	–2	1

(1,1) is the solution.

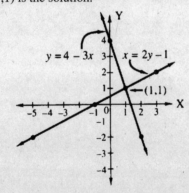

5. $3x - 4y = 18$

x	6	2	–2
y	0	–3	–6

$2y = -3x$

x	0	2	–2
y	0	–3	3

(2,–3) is the solution.

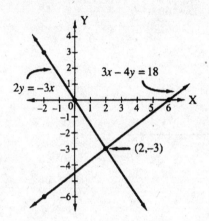

7. $5x - 2y = 20$

x	4	2	6
y	0	–5	5

(6,5) is the solution.

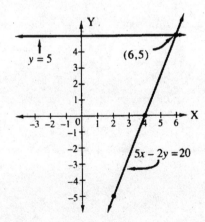

B. 1. For $y - 4x = 0$

x	0	5000	10 000
y	0	20 000	40 000

For $y - 2x - 10\,000 = 0$

x	0	5000	10 000
y	10 000	20 000	30 000

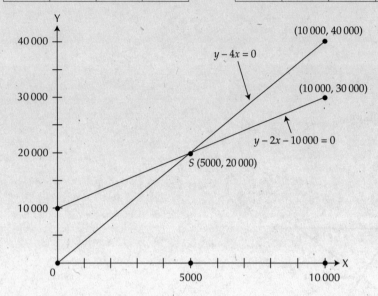

3. For $3x + 3y = 2400$

x	0	400	800
y	800	400	0

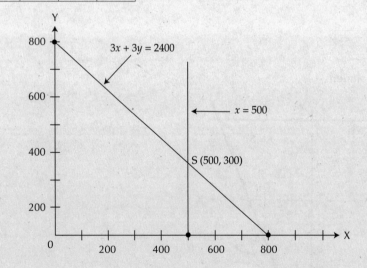

Exercise **4.4**

A. 1. Let the number of units of Product A be x and the number of units of Product B be y.

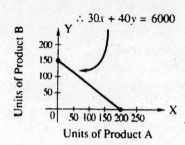

∴ $30x + 40y = 6000$

3. Let the number of tax returns be x and the total cost be $\$y$.

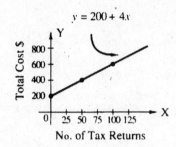

$y = 200 + 4x$

B. 1. Let the larger number be x and the smaller number be y.

$$\therefore x + y = 24 \qquad ①$$
$$2x = 3y + 3 \qquad ②$$

$$① \times 3 \to \quad 3x + 3y = 72$$
$$2x - 3y = 3$$
Add: $\qquad\qquad 5x = 75$
$$x = 15$$
$$y = 9$$

The numbers are 15 and 9.
Check:
Sum $= 15 + 9 = 24$
$2 \times 15 = (3 \times 9) + 3$

3. Let the number of jars of Brand X be x and the number of jars of the No-Name brand be y
$\therefore x + y = 140$ ①
$2.25x + 1.75y = 290$ ②

② × 4 → $9x + 7y = 1160$
① × 7 → $7x + 7y = 980$
Subtract: $2x = 180$
$x = 90$
$y = 50$

Sales were 90 jars of Brand X and 50 jars of No-Name brand.
Check:
Total sold $= 90 + 50 = 140$
Value: $90 \times 2.25 = \$202.50$
$50 \times 1.75 = \underline{\quad 87.50}$
$= \$290.00$

5. Let Kaya's investment be $\$x$, and Fred's investment be $\$y$.
$\therefore x + y = 55\,000$ ①
$y = \dfrac{2}{3}x + 2500$ ②

② × 3 → $-2x + 3y = 7500$
① × ② → $2x + 2y = 110\,000$
Add: $5y = 117\,500$
$y = 23\,500$
$x = 31\,500$

Kaya's investment is $\$31\,500$ and Fred's investment is $\$23\,500$.
Check:
Total $= 31\,500 + 23\,500 = \$55\,000$
$\dfrac{2}{3} \times 31\,500 + 2500 = 21\,000 + 2500 = \$23\,500$

7. Let the number of Type A lights be x and the number of Type B lights be y.
$\therefore x + y = 60$ ①
$40x + 50y = 2580$ ②

② ÷ 10 → $4x + 5y = 258$
① × (−4) → $-4x - 4y = -240$
Add: $y = 18$
$x = 42$

The number of Type A is 42, and the number of Type B is 18.
Check:
Total $= 42 + 18 = 60$
Value:
$42 \times 40 = \$1680$
$18 \times 50 = \underline{\quad 900}$
$\$2580$

9.   Let the number of quarters be x and the number of loonies be y.

$$\therefore \quad 25x + 100y = 8575 \quad \text{①}$$

② →

$$x = \frac{3}{4}y + 1 \quad \text{②}$$

①. ÷ 25

$$x + 4y = 343 \quad \text{③}$$

② × 4

$$4x = 3y + 4 \quad \text{④}$$

③ × 4

$$4x + 16y = 1372 \quad \text{⑤}$$

Rearrange ④

$$4x - 3y = 4 \quad \text{⑥}$$

⑤ − ⑥

$$19y = 1368$$
$$y = 72$$

Substitute ②

$$x = \frac{3}{4}(72) + 1 = 54 + 1 = 55$$

Marysia has 55 quarters and 72 loonies.

Value:

$$
\begin{array}{rl}
55 \times 0.25 &= \$13.75 \\
72 \times 1.00 &= \underline{72.00} \\
&\ \ \$85.75
\end{array}
$$

Review Exercise

1. (a)   For $7x + 3y = 6$

$$3y = -7x + 6$$
$$y = \frac{-7x}{3} + 2$$

Slope, $m = -\dfrac{7}{3}$; y-intercept, $b = 2$

(b)   For $10y = 5x$

$$y = \frac{5x}{10}$$
$$y = \frac{1}{2}x$$

Slope, $m = \dfrac{1}{2}$; y-intercept, $b = 0$

(c)   For $\dfrac{2y - 3x}{2} = 4$

$$2y - 3x = 8$$
$$2y = 3x + 8$$
$$y = \frac{3}{2}x + 4$$

Slope, $m = \dfrac{3}{2}$; y-intercept, $b = 4$

(d)   For $1.8x + 0.3y - 3 = 0$

$$18x + 3y - 30 = 0$$
$$3y = -18x + 30$$
$$y = -6x + 10$$

Slope, $m = -6$; y-intercept, $b = 10$

(e)   For $\dfrac{1}{3}x = -2$

$$x = -6$$

Line is parallel to the y-axis.
Slope, m is undefined.
There is no y-intercept.

(f)   For $11x - 33y = 99$

$$-33y = -11x + 99$$
$$y = \frac{-11x}{-33} + \frac{99}{-33}$$
$$y = \frac{1}{3}x - 3$$

Slope, $m = \dfrac{1}{3}$; y-intercept, $b = -3$

(g)   For $xy - (x + 4)(y - 1) = 8$

$$xy - xy - 4y + x + 4 = 8$$
$$-4y = -x + 4$$
$$y = \frac{1}{4}x - 1$$

Slope, $m = \dfrac{1}{4}$; y-intercept, $b = -1$

(h)   For $2.5y - 12.5 = 0$

$$25y - 125 = 0$$
$$y = 5$$

Line is parallel to the x-axis;
slope, $m = 0$; y-intercept, $b = 5$

3. (a) $3x + y = 6$ and $x - y = 2$

x	0	2	4
y	6	0	-6

x	0	2	4
y	-2	0	2

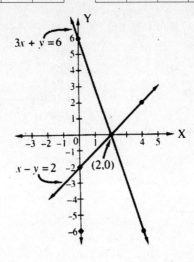

(b) $x + 4y = -8$ and $3x + 4y = 0$

x	0	4	-4
y	-2	-3	-1

x	0	4	-4
y	0	-3	3

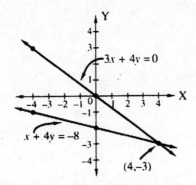

71

(c) $5x = 3y$ and $y = -5$

x	0	3	−3
y	0	5	−5

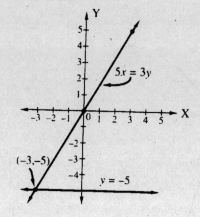

(d) $2x + 6y = 8$ and $x = -2$

x	4	−2	1
y	0	2	1

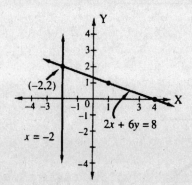

(e) $y = 3x - 2$ and $y = 3$

x	0	$5/3$	–1
y	–2	3	–5

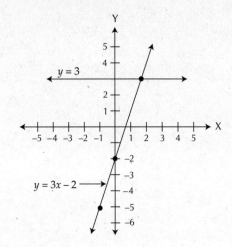

(f) $y = -2x$ and $x = 4$

x	0	4	–2
y	0	–8	4

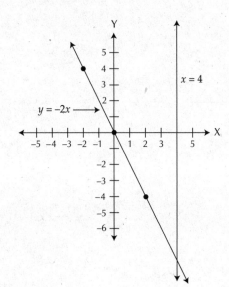

(g) $x = -2$, $y = 0$, and $3x + 4y = 12$

x	0	−2	4
y	3	4.5	0

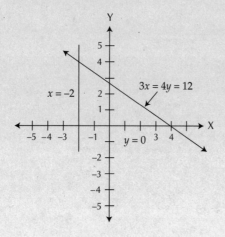

(h) $x = 0$, $y = -2$ and $5x + 3y = 15$

x	3	0	4.2
y	0	5	−2

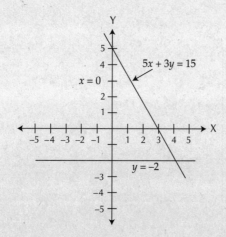

5.　(a)　Let the number of announcements be x and the total cost be $\$y$.

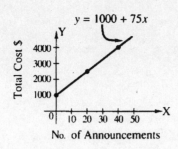

(b)　Let the number of units of Product A be x and the number of units of Product B be y.

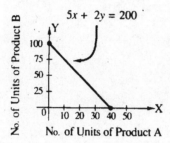

Self-Test

1.　(a)　For $4y + 11 = y$
$$y = -\frac{11}{3}$$
Slope, $m = 0$; y-intercept, $b = -\frac{11}{3}$

(b)　For $\frac{2}{3}x - \frac{1}{9}y = 1$
$$6x - y = 9$$
$$y = 6x - 9$$
Slope, $m = 6$; y-intercept, $b = -9$

(c)　For $x + 3y = 0$
$$3y = -x$$
$$y = -\frac{1}{3}x$$
Slope, $m = -\frac{1}{3}$; y-intercept, $b = 0$

(d) For $-6y - 18 = 0$

$$-6y = 18$$
$$y = -3$$

Line is parallel to the x-axis;
slope, $m = 0$; y-intercept, $b = -3$

(e) For $13 - \dfrac{1}{2}x = 0$

$$26 - x = 0$$
$$x = 26$$

Line is parallel to the y-axis;
slope, m is undefined; there is no y-intercept

(f) For $ax + by = c$

$$by = -ax + c$$
$$y = -\dfrac{a}{b}x + \dfrac{c}{b}$$

Slope, $m = -\dfrac{a}{b}$; y-intercept, $b = \dfrac{c}{b}$

3. (a) For $-x = -55 + y$

x	0	25	55
y	55	30	0

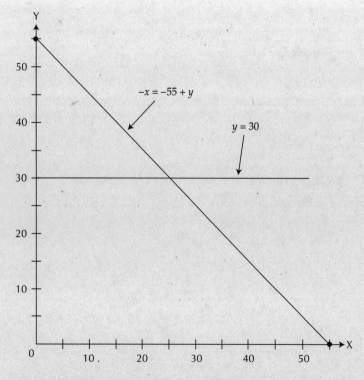

(b)   For $3x + 2y - 600 = 0$

x	0	100	200
y	−300	−450	−600

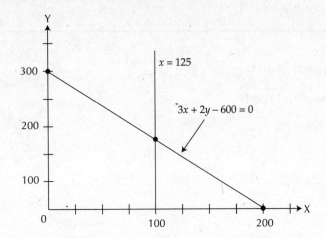

5.   Let the amount invested at 8% be x, and the amount invested at 12% be y.

$x + y = 12\,000$ ①

The amount of interest earned at 8% is $0.08x$; the amount of interest earned at 12% is $0.12y$.

$0.08x + 0.12y = 1120$ ②

To eliminate x from ① and ②

② × 100 → $8x + 12y = 112\,000$
① × 8 → $8x + 8y = 96\,000$
Subtract: $4y = 16\,000$
                         $y = 4000$
In ① $x + 4000 = 12\,000$
                  $x = 8000$

The amount invested at 8% is $8000; the amount invested at 12% is $4000.

PART TWO   *Mathematics of business and management*

5 Business applications—Depreciation and break-even analysis

Exercise **5.1**

A. 1. (a)   Original book value $40 000
            Scrap value   5 000
            Wearing value $35 000

   Annual depreciation $= \dfrac{35\,000}{10} = \$3500$

Depreciation Schedule—Straight Line Method

End of year	Depreciation expense	Accumulated depreciation	Book value
0			40 000
1	3 500	3 500	36 500
2	3 500	7 000	33 000
3	3 500	10 500	29 500
4	3 500	14 000	26 000
5	3 500	17 500	22 500
6	3 500	21 000	19 000
7	3 500	24 500	15 500
8	3 500	28 000	12 000
9	3 500	31 500	8 500
10	3 500	35 000	5 000
Total	35 000		

(b) Number of service hours $= 4(2025) + 3(1990) + 3(1775)$
$= 8100 + 5970 + 5325$
$= 19\ 395$

Wearing value $= 35\ 000$

Depreciation per service hour $= \dfrac{35\ 000}{19\ 395} = \$1.80$

Depreciation: Year 1 to 4 $= 2025 \times 1.80 = \$3645.00$
Year 5 to 7 $= 1990 \times 1.80 = \$3582.00$
Year 8 to 10 $= 1775 \times 1.80 = \$3195.00$

Depreciation Schedule—Service Hours Method

End of year	Depreciation expense	Accumulated depreciation	Book value
0			40 000.00
1	3 645.00	3 645.00	36 355.00
2	3 645.00	7 290.00	32 710.00
3	3 645.00	10 935.00	29 065.00
4	3 645.00	14 580.00	25 420.00
5	3 582.00	18 162.00	21 838.00
6	3 582.00	21 744.00	18 256.00
7	3 582.00	25 326.00	14 674.00
8	3 195.00	28 521.00	11 479.00
9	3 195.00	31 716.00	8 284.00
10	3 195.00	34 911.00	5 089.00
Total	34 911.00		

Note: Totals are affected by roudning.

(c) Number of units $= 4(10\ 750) + 3(9800) + 3(9200)$
$= 43\ 000 + 29\ 400 + 27\ 600$
$= 100\ 000$

Wearing value $= \$35\ 000$

Depreciation per unit $= \dfrac{35\ 000}{100\ 000} = \$0.35$

Depreciation Schedule—Units of Product Method

End of year	Depreciation expense	Accumulated depreciation	Book value
0			40 000.00
1	3 762.50	3 762.50	36 237.50
2	3 762.50	7 525.00	32 475.00
3	3 762.50	11 287.50	28 712.50
4	3 762.50	15 050.00	24 950.00
5	3 430.00	18 480.00	21 520.00
6	3 430.00	21 910.00	18 090.00
7	3 430.00	25 340.00	14 660.00
8	3 220.00	28 560.00	11 440.00
9	3 220.00	31 780.00	8 220.00
10	3 220.00	35 000.00	5 000.00
Total	35 000.00		

(d) Wearing value = \$35 000; $n = 10$

$10k + 9k + 8k + \ldots + k = 35\,000$

$$\frac{1}{2}(10)(11)k = 35\,000$$

$$55k = 35\,000$$

$$k = 636.3636$$

Depreciation Schedule—Sum of Years Digits Method

End of year	Proportional parts	Depreciation expense	Accumulated depreciation	Book value
0				40 000.00
1	10	6 363.64	6 363.64	33 636.36
2	9	5 727.27	12 090.91	27 909.09
3	8	5 090.91	17 181.82	22 818.18
4	7	4 454.55	21 636.37	18 363.63
5	6	3 818.18	25 454.55	14 545.45
6	5	3 181.82	28 636.37	11 363.63
7	4	2 545.45	31 181.82	8 818.18
8	3	1 909.10	33 090.92	6 909.08
9	2	1 272.73	34 363.65	5 636.35
10	1	636.35	35 000.00	5 000.00
Total	55	35 000.00		

(e) Rate of depreciation $= 2 \times \dfrac{1}{10} = 0.20 = 20\%$

Depreciation Schedule—Simple Declining Balance Method

End of year	Depreciation expense	Accumulated depreciation	Book value
0			40 000.00
1	8 000.00	8 000.00	32 000.00
2	6 400.00	14 400.00	25 600.00
3	5 120.00	19 520.00	20 480.00
4	4 096.00	23 616.00	16 384.00
5	3 276.80	26 892.80	13 107.20
6	2 621.44	29 514.24	10 485.76
7	2 097.15	31 611.39	8 388.61
8	1 677.72	33 289.11	6 710.89
9	1 342.18	34 631.29	5 368.71
10	368.71	35 000.00	5 000.00
Total	35 000.00		

(f) Original cost $= 40\,000.00$; residual value $= 5000.00$; $n = 10$

$$d = 1 - \sqrt[10]{\frac{5000}{40\,000}} = 1 - 0.125^{\frac{1}{10}} = 1 - 0.8122524 = 18.77476\%$$

Depreciation Schedule—Complex Declining Balance Method

End of year	Depreciation expense	Accumulated depreciation	Book value
0			40 000.00
1	7 509.90	7 509.90	32 490.10
2	6 099.94	13 609.84	26 390.16
3	4 954.69	18 564.53	21 435.47
4	4 024.46	22 588.99	17 411.01
5	3 268.87	25 857.86	14 142.14
6	2 655.16	28 513.02	11 486.98
7	2 156.65	30 669.67	9 330.33
8	1 751.75	32 421.42	7 578.58
9	1 422.86	33 844.28	6 155.72
10	1 155.72	35 000.00	5 000.00
Total	35 000.00		

(g) Class 8 rate $= 20\%$

Capital Cost Allowance Schedule

End of year	Capital cost allowance	Accumulated capital cost allowance	Book value
0			40 000.00
1	8 000.00	8 000.00	32 000.00
2	6 400.00	14 400.00	25 600.00
3	5 120.00	19 520.00	20 480.00
4	4 096.00	23 616.00	16 384.00
5	3 276.80	26 892.80	13 107.20
6	2 621.44	29 514.24	10 485.76
7	2 097.15	31 611.39	8 388.61
8	1 677.72	33 289.11	6 710.89
9	1 342.18	34 631.29	5 368.71
10	1 073.74	35 705.03	4 294.97
Total	35 705.03		

2.   (a)   Original cost $= \$32\,000$; residual value $= \$5000$; $n = 8$
         Wearing value $= 32\,000 - 5000 = \$27\,000$

         Yearly depreciation $= \dfrac{1}{8}(27\,000) = \$3375$

Depreciation Schedule—Straight Line Method

End of year	Depreciation expense	Accumulated depreciation	Book value
0			32 000
1	3 375	3 375	28 625
2	3 375	6 750	25 250
3	3 375	10 125	21 875
4	3 375	13 500	18 500
5	3 375	16 875	15 125
6	3 375	20 250	11 750
7	3 375	23 625	8 375
8	3 375	27 000	5 000
Total	27 000		

  (b)   Wearing value $= \$27\,000$; $n = 8$
         $8k + 7k + 6k + \dots + k = 27\,000$

$$\frac{1}{2}(8)(9)k = 27\,000$$

$$36k = 27\,000$$

$$k = 750$$

Depreciation Schedule—Sum of Years Digit Method

End of year	Proportional parts	Depreciation expense	Accumulated depreciation	Book value
0				32 000
1	8	6 000	6 000	26 000
2	7	5 250	11 250	20 750
3	6	4 500	15 750	16 250
4	5	3 750	19 500	12 500
5	4	3 000	22 500	9 500
6	3	2 250	24 750	7 250
7	2	1 500	26 250	5 750
8	1	750	27 000	5 000
Total	36	27 000		

(c)   Rate of depreciation $= 2 \times \dfrac{1}{8} = 0.25 = 25\%$

Depreciation Schedule—Simple Declining Balance Method

End of year	Depreciation expense	Accumulated depreciation	Book value
0			32 000.00
1	8 000.00	8 000.00	24 000.00
2	6 000.00	14 000.00	18 000.00
3	4 500.00	18 500.00	13 500.00
4	3 375.00	21 875.00	10 125.00
5	2 531.25	24 406.25	7 593.75
6	1 898.44	26 304.69	5 695.31
7	695.31	27 000.00	5 000.00
8	—	27 000.00	5 000.00
Total	27 000.00		

(d)   Original cost $= \$32\,000$; residual value $= \$5000$; $n = 8$

$$d = 1 - \sqrt[8]{\frac{5000}{32\,000}} = 1 - 0.15625^{\frac{1}{8}} = 1 - 0.7929166 = 20.70834\%$$

Depreciation Schedule—Complex Declining Balance Method

End of year	Depreciation expense	Accumulated depreciation	Book value
0			32 000.00
1	6 626.67	6 626.67	25 373.33
2	5 254.40	11 881.07	20 118.93
3	4 166.29	16 047.36	15 952.64
4	3 303.53	19 350.89	12 649.11
5	2 619.42	21 970.31	10 029.69
6	2 076.98	24 047.29	7 952.71
7	1 646.88	25 694.17	6 305.83
8	1 305.83	27 000.00	5 000.00
Total	27 000.00		

B.   1.   (a)   By the straight line method:

Original cost $= \$560\,000$
Salvage value $= \underline{\quad 68\,000}$
   Wearing value $= \$492\,000$

Yearly depreciation $= \dfrac{492\,000}{40} = \$12\,300$

(i)   Accumulated depreciation after 15 years $= 12\,300(15) = \$184\,500$
Book value $= 560\,000 - 184\,500 = \$375\,500$

(ii) Depreciation charge in year 16 $= \$12\,300$

(b) By the sum-of-the-years-digits method:
Wearing value $= 492\,000$
$$40k + 39k + 38k + \ldots + k = 492\,000$$
$$\frac{1}{2}(40)(41)k = 492\,000$$
$$820k = 492\,000$$
$$k = \$600$$

(i) Accumulated depreciation after 15 years:

Sum of parts for all 40 years $= \dfrac{1}{2}(40)(41) \qquad = 820$

Sum of parts for the last 25 years $= \dfrac{1}{2}(25)(26) = \underline{325}$

Sum of parts assigned to the first 15 years $\qquad = 495$
Accumulated depreciation $= 495(600) = \$297\,000$
Book value after 15 years $= 560\,000 - 297\,000 = \$263\,000$

(ii) Since Year 16 is the first of the remaining 25 years, the number of parts assigned to year 16 is 25.
Depreciation in Year 16 $= 25(600) = \$15\,000$

(c) Simple declining balance method:
$$\text{Rate} = 2 \times \frac{1}{40} = \frac{1}{20} = 0.05 = 5\%$$

(i) Book value after 15 years $= 560\,000(0.95)^{15}$
$$= 560\,000(0.4632912)$$
$$= \$259\,443.07$$

(ii) Depreciation in Year 16 $= 259\,443.07(0.05)$
$$= \$12\,972.15$$

(d) Complex declining balance method:
$$\text{Rate} = 1 - \sqrt[40]{\frac{68\,000}{560\,000}} = 1 - 0.1214286^{\frac{1}{40}} = 1 - 0.9486544 = 5.13456\%$$

(i) Book value after 15 years
$$= 560\,000\left(0.9486544^{15}\right)$$
$$= 560\,000(0.453545)$$
$$= \$253\,985.20$$

(ii) Depreciation in Year 16
$$= 253\,985.20(0.0513456)$$
$$= \$13\,041.02$$

2.   (a)   Straight line method:

Original cost   = $150 000

Scrap value   = <u> 13 500</u>

   Wearing value   = $136 500

Yearly depreciation $= \dfrac{136\,500}{25} = \$5460$

  (i)   Accumulated depreciation after 10 years

       $= 10(5460) = \$54\,600$

       Book value $= 150\,000 - 54\,600 = \$95\,400$

  (ii)   Depreciation in Year 11 $= \$5460$

  (b)   Sum-of-the-years-digits method:

Wearing value $= 136\,500$

$25k + 24k + 23k + \ldots + k = 136\,500$

$$\tfrac{1}{2}(25)(26)k = 136\,500$$

$$325k = 136\,500$$

$$k = \$420$$

  (i)   Accumulated depreciation after 10 years:

Sum of parts for all 25 years $= \tfrac{1}{2}(25)(26)$ $= 325$

Sum of parts for the last 15 years $= \tfrac{1}{2}(15)(16)$ $= \underline{120}$

Sum of parts assigned to the first 10 years $= 205$

Accumulated depreciation after 10 years

$= 205(420) = \$86\,100$

Book value $= 150\,000 - 86\,100 = \$63\,900$

  (ii)   Since Year 11 is the first of the remaining 15 years, the number of parts assigned is 15.

Depreciation in Year 15 $= 15(420) = \$6300$

  (c)   Simple declining balance method:

Rate $= 2 \times \dfrac{1}{25} = 0.08 = 8\%$

  (i)   Book value after 10 years $= 150\,000(0.92)^{10}$

                 $= 150\,000(0.4343885)$

                 $= \$65\,158.28$

  (ii)   Depreciation in Year 11 $= 65\,158.28(0.08)$

                 $= \$5212.66$

(d)   Complex declining balance method:

$$\text{Rate} = 1 - \sqrt[25]{\frac{13\,500}{150\,000}} = 1 - 0.09^{\frac{1}{25}} = 1 - 0.9081753 = 9.18247\%$$

(i)   Book value after 10 years

$= 150\,000.00(0.9081753)^{10}$

$= 150\,000.00(0.3816778)$

$= \$57\,251.67$

(ii) Depreciation in Year 11

$= 57\,251.67(0.0918247)$

$= \$5257.12$

Exercise **5.2**

A. 1. (a)   (i) Revenue $= 120x$ $\Big\}$ where x represents the

(ii) Cost $= 2800 + 50x$ number of units per period.

(c)   (i) Contribution margin per unit $= 120.00 - (35.00 + 15.00) = \$70.00$

(ii) Contribution rate $= \dfrac{70}{120} = 0.5833333 = 58.33\%$

(d)   (i) To break even, revenue $=$ total cost

$$120x = 2800 + 50x$$
$$70x = 2800$$
$$x = 40$$

Break-even volume is 40 units.

(ii) B-E volume as a percent of capacity

$= \dfrac{40}{100} = 40\%$

(iii) B-E volume in sales dollars

$= 40 \times 120 = \$4800$

3. (a) Let x represent the sales volume in dollars.

   (i) Revenue $= x$

   (ii) Variable cost $= \dfrac{324\,000}{720\,000} = 0.45 = 45\%$ of revenue

   Total cost $= 220\,000 + 0.45x$

(b)

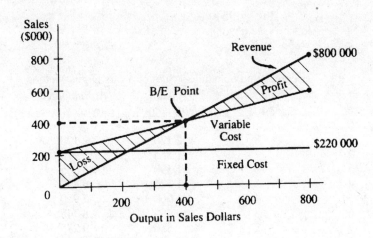

(c) (i) Contribution margin per dollar $= 1.00 - 0.45 = \$0.55$

   (ii) Contribution rate $= \dfrac{0.55}{1.00} = 0.55 = 55\%$

(d) (i) Not applicable

   (ii) B-E volume as a percent of capacity

   $= \dfrac{400\,000}{800\,000} = 50\%$

   (iii) To break even, $x = 220\,000 + 0.45x$

   $$0.55x = 220\,000$$
   $$x = 400\,000$$

   B-E volume is $\$400\,000$.

Review Exercise

1. (a) Original cost   = $12 000
        Trade-in value   =   2 460
          Wearing value = $ 9 540

$$8k \ + \ 7k \ + \ 6k \ + \ \dots \ + \ k \ = \ 9540$$

$$\frac{1}{2}(8)(9)k \ = \ 9540$$

$$36k \ = \ 9540$$

$$k \ = \ 265$$

Depreciation Schedule—Sum of Years Digits Method

End of year	Proportional parts	Annual depreciation	Accumulated depreciation	Book value
0				12 000
1	8	2 120	2 120	9 880
2	7	1 855	3 975	8 025
3	6	1 590	5 565	6 435
4	5	1 325	6 890	5 110
5	4	1 060	7 950	4 050
6	3	795	8 745	3 255
7	2	530	9 275	2 725
8	1	265	9 540	2 460
Total	36	9 540		

(b) Rate of depreciation $= \ 2 \times \dfrac{1}{8} \ = \ 0.25 \ = \ 25\%$

Depreciation Schedule—Simple Declining Balance Method

End of year	Annual depreciation	Accumulated depreciation	Book value
0			12 000.00
1	3 000.00	3 000.00	9 000.00
2	2 250.00	5 250.00	6 750.00
3	1 687.50	6 937.50	5 062.50
4	1 265.62	8 203.12	3 796.88
5	949.22	9 152.34	2 847.66
6	387.66	9 540.00	2 460.00
7	—	9 540.00	2 460.00
8	—	9 540.00	2 460.00
Total	9 540.00		

(c)   Rate of depreciation $= 1 - \sqrt[8]{\dfrac{2460}{12\,000}} = 1 - \left(\dfrac{2460}{12\,000}\right)^{\frac{1}{8}}$

$\qquad\qquad\qquad\qquad = 1 - 0.8202934 = 17.97066\%$

Depreciation Schedule—Complex Declining Balance Method

End of year	Depreciation expense	Accumulated depreciation	Book value
0			12 000.00
1	2 156.48	2 156.48	9 843.52
2	1 768.94	3 925.42	8 074.58
3	1 451.06	5 376.48	6 623.52
4	1 190.29	6 566.77	5 433.23
5	976.39	7 543.16	4 456.84
6	800.92	8 344.08	3 655.92
7	656.99	9 001.07	2 998.93
8	538.93	9 540.00	2 460.00
Total	9 540.00		

3.   Original cost $\quad= \$9160$
  Trade-in $\qquad\quad= \underline{2230}$
     Wearing value $\;= \$6930$
  $n = 6$
  $6k + 5k + 4k + 3k + 2k + k = 6930$

  $$\tfrac{1}{2}(6)(7)k = 6930$$
  $$21k = 6930$$
  $$k = 330$$

Depreciation Schedule—Sum of the Years Digits Method

End of year	Proportional parts	Annual depreciation	Accumulated depreciation	Book value
0				9 160
1	6	1 980	1 980	7 180
2	5	1 650	3 630	5 530
3	4	1 320	4 950	4 210
4	3	990	5 940	3 220
5	2	660	6 600	2 560
6	1	330	6 930	2 230
Total	21	6 930		

5.   Original cost $= \$75\,000$; $n = 20$; residual value $= \$3500$

  Rate of depreciation $= 1 - \sqrt[20]{\dfrac{3500}{75\,000}} = 1 - 0.0466667^{\frac{1}{20}} = 1 - 0.857927 = 14.2073\%$

  Book value after 11 years
  $= 75\,000\left(0.857927^{11}\right) = 75\,000(0.1853334) = \$13\,900.01$

  Depreciation in Year 12 $= 13\,900.01(0.142073) = \$1974.82$

7. (a)   Let x represent the sales volume in dollars.

Revenue $= x$

Variable cost $= \dfrac{13\,552}{19\,360} = 0.70x$

Total cost $= 4800 + 0.70x$

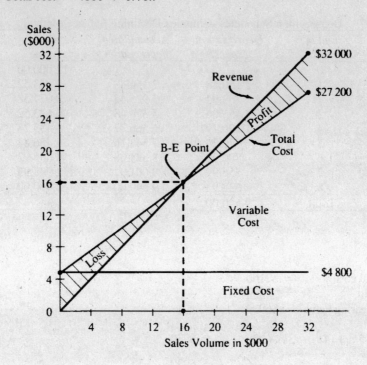

(b)   (i)   Contribution margin per dollar $= 1.00 - 0.70 = \$0.30$

(ii)   Contribution rate $= \dfrac{0.30}{1.00} = 0.30 = 30\%$

(c)   (i)
$$x = 4800 + 0.70x$$
$$0.30x = 4800$$
$$x = 16\,000$$
B.E. volume is $16 000

(ii)   B-E point as a percent of capacity
$$= \dfrac{16\,000}{32\,000}$$
$$= 50\%$$

(d) (i)   FC $= 4800 - 600 = 4200$

$$x = 4200 + 0.70x$$
$$0.30x = 4200$$
$$x = 14\ 000$$

B-E point is a sales volume of \$14 000,

or at $\dfrac{14\ 000}{32\ 000} = 43.75\%$ of capacity.

(ii)   VC $= 0.55x$

$$x = 5670 + 0.55x$$
$$0.45x = 5670$$
$$x = 12\ 600$$

B-E point is a sales volume of \$12 600,

or at $\dfrac{12\ 600}{32\ 000} = 39.375\%$ of capacity.

9. (a) (i) Contribution margin per unit $= 640.00 - 360.00 = \$280.00$

      (ii) Contribution rate $= \dfrac{280.00}{640.00} = 0.4375 = 43.75\%$

(b) Let the number of units be x.

Revenue $= 640x$

Total cost $= 26\ 880 + 360x$

(i) To break even

$$640x = 26\ 880 + 360x$$
$$280x = 26\ 880$$
$$x = 96$$

The B-E point is 96 units.

(ii) B-E point as a percent of capacity

$$= \dfrac{96}{150} = 0.64 = 64\%$$

(iii) B-E point in dollars

$$= 96(640) = \$61\ 440$$

(c) For FC $= \$32\ 200$

$$640x = 32\ 200 + 360x$$
$$280x = 32\ 200$$
$$x = 115$$

B-E point in dollars is $115(640) = \$73\ 600$.

(d)   VC $= 60\%$ of $640x = 384x$

$$640x = 23\ 808 + 384x$$
$$256x = 23\ 808$$
$$x = 93$$

B-E point $= \dfrac{93}{150} = 0.62 = 62\%$ of capacity.

1. Original book value = \$85 000; salvage value = \$5000;
   Wearing value = \$80 000; $n = 16$

   Annual depreciation $= \dfrac{80\,000}{16} = \$5000$

   Accumulated depreciation after 10 years
   $= 10 \times 5000 = \$50\,000$
   Book value after ten years $= 85\,000 - 50\,000 = \$35\,000$

3. Original cost = \$45 000; residual value = \$2500;
   $n = 20$

   Rate of depreciation $= 1 - \sqrt[20]{\dfrac{2500}{45\,000}}$

   $= 1 - 0.0555556^{\frac{1}{20}}$
   $= 1 - 0.8654388 = 13.45612\%$

   Book value after 14 years
   $= 45\,000(0.8654388)^{14} = 45\,000(0.1322236) = \$5950.06$
   Depreciation in Year 15 $= 5950.06(0.1345612) = \$800.65$

5. (a) Let Revenue be x.

   $\text{Variable cost} = \dfrac{270\,000}{600\,000} \times \text{Revenue} = 0.45x$

   $\text{Total cost} = 275\,000 + 0.45x$

      (i) Contribution margin per dollar = 1.00 - 0.45 = \$0.55

      (ii) Contribution rate $= \dfrac{0.55}{1.00} = 55\%$

   (b) (i) $x = 275\,000 + 0.45x + \text{NI}$
   To break even, NI = 0
   $x = 275\,000 + 0.45x$
   $0.55x = 275\,000$
   $x = \$500\,000$

      (ii) B-E volume as a percent of capacity

   $= \dfrac{500\,000}{800\,000} = 0.625 = 62.5\%$

   (c) $x = (275\,000 + 40\,000) + 0.40x$
   $0.60x = 315\,000$
   $x = \$525\,000$

6 Commercial discount, markup and markdown

Exercise **6.1**

A. 1. Discount = Rate × List
       Discount = 45% × 24.60 = $11.07
         Net = NCF × List
             = (1 − 0.45)(24.60)
             = (0.55)(24.60)
             = $13.53

3. Discount = 76.95 − 51.30 = $25.65

$$\text{Rate} = \frac{\$\,\text{Discount}}{\text{List}} = \frac{25.65}{76.95} = 33\frac{1}{3}\%$$

5. List = 214.71 + 37.89 = $252.60

$$\text{Rate} = \frac{37.89}{252.60} = 15\%$$

7.
$$\text{List} = \frac{\$\,\text{Discount}}{\text{Rate}} = \frac{83.35}{0.625} = \$133.36$$

Net = 133.36 − 83.35 = $50.01

9.   84.35 = 0.625L

$$L = \frac{84.35}{0.625}$$

List = $134.96

Discount = 134.96 − 84.35 = $50.61

B. 1. Discount = 975.00 − 820.00 = $155.00

$$\text{Rate} = \frac{155.00}{975.00} = 15.9\%$$

3. 37.5% of List = 913.50

    $$0.375L = 913.50$$

    $$\text{List} = \frac{913.50}{0.375} = \$2436.00$$

5. $16\frac{2}{3}\%$ of List = 14.82

    $$\frac{1}{6}L = 14.82$$

    $$L = 88.92$$

    Net = 88.92 − 14.82 = \$74.10

7. $$0.8\dot{3}L = 355.00$$

    $$L = \frac{355.00}{0.8\dot{3}}$$

    List = \$426.00

9. Net = 0.80(85.00) = \$68.00

    Reduction needed = 68.00 − 57.80 = \$10.20

    Additional discount rate $= \dfrac{10.20}{68.00} = 15\%$

Exercise 6.2

A. 1. Net = (0.75)(0.90)(44.80) = \$30.24

    Single rate $= \dfrac{44.80 - 30.24}{44.80} = 0.325 = 32.5\%$

    3. Net = (0.60)(0.875)(0.98)(268.00) = \$137.89

    Single rate = 1 − (0.60)(0.875)(0.98) = 1 − 0.5145 = 0.4855 = 48.55%

    5. $617.50 = (0.65)(0.6\dot{6})(0.90)L$

    $$L = \frac{617.50}{(0.65)(0.6\dot{6})(0.90)} = \$1583.33$$

    Single rate = 1 − (0.65)(0.6\dot{6})(0.90) = 1 − 0.39 = 0.61 = 61%

B. 1. (a) Net = (0.70)(0.80)(0.95)(240.00) = \$127.68

(b) Discount = 240.00 − 127.68 = \$112.32

(c) Single rate = $\frac{112.32}{240.00}$ = 46.8%

3. (a) Single rate = 1 − (0.70)(0.875) = 1 − 0.6125 = 0.3875 = 38.75%

(b) Single rate = 1 − $(0.6\dot{6})$(0.80)(0.97) = 1 − 0.5173 = 0.4826 = 48.26%

5. Net = 750.00(0.80)(0.95)(0.98) = \$558.60
Additional discount = 558.60 − 474.81 = \$83.79
Additional rate = $\frac{83.79}{558.60}$ = 15%

7. 113.40 = (0.75)(0.875)(0.96)L
$$L = \frac{113.40}{(0.75)(0.875)(0.96)}$$
List = \$180.00

Exercise 6.3

A. 1. No discount allowed
Pay \$640.00

3. Allow 1% discount
Pay = (0.99)(783.95) = \$776.11

5. Allow 2% discount
Pay = (0.98)(1160.00) = \$1136.80

7.   Allow 2% discount
     Pay $= (0.98)(4675.00) = \$4581.50$

B.   1.   Net $= (0.97)(600) = \$582.00$
          Balance due $= 1450.00 - 600.00 = \$850.00$

     3.   Credit $= 964.50 - 400.00 = \$564.50$
          Net received $= (0.95)(564.50) = \$536.28$

     5. $785.70 = 0.97$ credit
          Credit $= \dfrac{785.70}{0.97} \doteq \$810.00$

          Balance due $= 1620.00 - 810.00 = \$810.00$

C.   1.   (a)   September 10

          (b) $5(980)(0.75)(0.95)$ $3491.25
                $4(696)\left(0.8\overline{3}\right)(0.875)(0.96)$ 1948.80
                          Amount of invoice $5440.05
                  Less: Discount of 3%   163.20
                              Amount due $5276.85

          (c)   Partial payment credit $= 5440.05 - 2000.00 = \$3440.05$
                Cash discount $= (0.03)(3440.05) = \$103.20$

     3.

Invoice date	Amount	Rate of discount	Net amount paid	
July 25	$929.00	—		$ 929.00
August 10	763.00	3%	(0.97)(763.00)	740.11
August 29	864.00	3%	(0.97)(864.00)	838.08
Amount remitted				$2507.19

     5.   Allow 4% discount on partial payment of $2275.00
          Amount paid $= (0.96)(2275.00) = \$2184.00$

7.　(a)　Allow discount of 3% on partial payment of $1200.00
　　　　Amount paid = (0.97)(1200.00) = $1164.00

　　(b)　Allow discount of 1% on partial payment of $740.95
　　　　Amount paid = (0.99)(740.95) = $733.54

　　(c)　Amount paid on October 30 is $600.00.

9.　(a)　Allow 3% on unknown partial payment $P
　　　　0.97P = 1867.25
　　　　　　P = 1925.00
　　　　Credit to the account is $1925.00.

　　(b)　Amount owing = 5325.00 − 1925.00 = $3400.00

Exercise **6.5**

A.　1.　(a)　30.00 − 24.00 = $6.00

　　　　(b)　(0.16)(24) = $3.84

　　　　(c)　6.00 − 3.84 = $2.16

　　　　(d)　$\dfrac{6}{24}$ = 0.25 = 25%

　　　　(e)　$\dfrac{6}{30}$ = 0.20 = 20%

　　3.　(a)　87.50 − 52.50 = $35.00

　　　　(b)　(0.36)(87.50) = $31.50

(c)   35.00 − 31.50 = $3.50

(d) $\dfrac{35.00}{52.50} = 0.6\dot{6} = 66\dfrac{2}{3}\%$

(e) $\dfrac{35.00}{87.50} = 0.40 = 40\%$

5.   (a)   37.50 − 27.00 = $10.50

(b)   (0.34)(37.50) = $12.75

(c)   10.50 − 12.75 = ($2.25)

(d) $\dfrac{10.50}{27.00} = 0.3\dot{8} = 38\dfrac{8}{9}\%$

(e $\dfrac{10.50}{37.50} = 0.28 = 28\%$

B.   1.   Markup = 31.25 − 25.00 = $6.25

Rate of markup based on cost = $\dfrac{6.25}{25.00} = 0.25 = 25\%$

Rate of markup based on selling price = $\dfrac{6.25}{31.25} = 0.20 = 20\%$

3.   Selling price = 64.00 + 38.40 = $102.40

Rate of markup based on C = $\dfrac{38.40}{64.00} = 0.60 = 60\%$

Rate of markup based on SP = $\dfrac{38.40}{102.40} = 0.375 = 37.5\%$

5. Markup $= 0.40 \times 54.25 = \$21.70$
   Selling price $= 54.25 + 21.70 = \$75.95$
   Rate of markup based on SP $= \dfrac{21.70}{75.95} = 0.2857143 = 28.6\%$

7. $C + M = S$
   $C + 50\% \text{ of } C = S$
   $C + 0.5C = 66.36$
   $1.5C = 66.36$
   $C = 44.24$
   Cost $= \$44.24$
   Markup $= 66.36 - 44.24 = \$22.12$
   Rate of markup based on SP $= \dfrac{22.12}{66.36} = 0.33 = 33\dfrac{1}{3}\%$

9. $C + 60\% \text{ of } S = S$
   $31.24 + 0.6S = S$
   $31.24 = 0.4S$
   $S = 78.10$
   Selling price is $78.10
   Markup $= (0.6)(78.10) = \$46.86$
   Rate of markup based on cost $= \dfrac{46.86}{31.24} = 1.50 = 150\%$

11. $16\dfrac{2}{3}\% \text{ of } S = 22.26$
    $\dfrac{1}{6}S = 22.26$
    $S = 133.56$
    Selling price $= \$133.56$
    Cost $= 133.56 - 22.26 = \$111.30$
    Rate of markup based on cost
    $= \dfrac{22.26}{111.30} = 0.20 = 20\%$

CHAPTER 6

C. 1. Cost per dozen
$$= (0.80)(0.80)(5.00)$$
$$= \$3.20$$
$$S = C + E + P$$
$$= C + 0.45C + 0.15C$$
$$= 1.60C$$
$$= (1.60)(3.20)$$
$$= \$5.12$$
Selling price should be \$5.12 per dozen.

3. (a) Clearance sale price $= (0.60)(125.00) = \$75.00$

(b) Inventory sale price
$$= \text{Cost} + \text{Overhead}$$
$$= C + 0.5C$$
$$= 1.5C = 1.5(42.00) = \$63.00$$

(c) Total revenue
$$= 120(125.00) + 60(75.00) + 20(63.00)$$
$$= 15\,000.00 + 4500.00 + 1260.00 = \$20\,760.00$$
Total cost
$$= 200(42.00) + 50\% \text{ of } (200.00 \times 42) = 8400.00 + 4200.00 = \$12\,600.00$$
Total profit $= 20\,760.00 - 12\,600.00 = \$8160.00$

(d) Average rate of markup
$$= \frac{\text{Total markup realized}}{\text{Total purchase cost}}$$
$$= \frac{20\,760.00 - 8400.00}{8400.00} = \frac{12\,360.00}{8400.00} = 1.4714286 = 147.14\%$$

5. Cost $= (0.60)(0.75)(55.00) = \$24.75$
Markup $= 54.45 - 24.75 = \$29.70$

(a) Markup based on cost $= \dfrac{29.70}{24.75} = 1.2 = 120\%$

(b) Markup based on selling price $= \dfrac{29.70}{54.45} = 0.54545 = 54.55\%$

7. (a)   15% of cost $= 3.42$

$$0.15C = 3.42$$
$$C = 22.80$$

Cost was $22.80

    (b)   Selling price $= 22.80 + 3.42 = \$26.22$

    (c)   Markup based on selling price $= \dfrac{3.42}{26.22} = 0.1304 = 13.04\%$

9. (a) $C + 40\%$ of $C = S$

$$C + 0.40C = 74.55$$
$$1.4C = 74.55$$
$$C = 53.25$$

Cost was $53.25

    (b)   Markup $= (0.40)(53.25) = \$21.30$

Rate of markup based on selling price $= \dfrac{21.30}{74.55} = 0.28571 = 28.57\%$

11. (a) $C + 60\%$ of $S = S$

$$12.80 + 0.60S = S$$
$$12.80 = 0.40S$$
$$S = 32.00$$

Selling price is $32.00

    (b)   Markup $= 32.00 - 12.80 = \$19.20$

Markup based on cost $= \dfrac{19.20}{12.80} = 1.50 = 150\%$

Exercise 6.6

A. 1. $51.00 + 25\% \text{ of } S = S$

$51.00 + 0.25S = S$

$51.00 = 0.75S$

$S = 68.00$

Sale price is $68.00

Regular selling price $-$ Discount $= S$

$R - 20\% \text{ of } R = 68.00$

$R - 0.20R = 68.00$

$0.80R = 68.00$

$R = 85.00$

Regular selling price is $85.00

Clearance sale price

$=$ Regular selling price $-$ Markdown

$= 85.00 - 26\% \text{ of } 85.00$

$= 85.00 - 22.10$

$= \$62.90$

3. $S =$ Regular selling price $-$ Discount

$= 144.00 - 16\frac{2}{3}\% \text{ of } 144.00$

$= 144.00 - 24.00 = 120.00$

Sale price is $120.00

Markup $= 120.00 - 40.00 = \$80.00$

Rate of markup based on cost $= \dfrac{80.00}{40.00} = 2.0 = 200\%$

Markdown $= 144.00 - 90.00 = \$54.00$

Rate of markdown $= \dfrac{54.00}{144.00} = 0.375 = 37.5\%$

5. $C + 60\% \text{ of } C = S$

$C + 0.60C = 72.80$

$1.6C = 72.80$

$C = 45.50$

Cost is $45.50

Clearance sale price $=$ Regular selling price $-$ Markdown

$61.60 = R - 36\% \text{ of } R$

$61.60 = 0.64R$

$R = 96.25$

Regular selling price is $96.25

Discount $= 96.25 - 72.80 = \$23.45$

Rate of discount $= \dfrac{23.45}{96.25} = 0.24364 = 24.36\%$

B.   1.   Sale price = Regular selling price − Markdown
                  = 85.00 − 40% of 85.00
                  = 85.00 − 34.00
                  = $51.00

Total cost = Cost + Expense
            = 42.00 + 20% of Regular selling price
            = 42.00 + 0.20(85.00)
            = 42.00 + 17.00
            = $59.00

Profit = Revenue − Total cost
        = 51.00 − 59.00
        = ⟨$8.00⟩

3.   Sale price = Regular selling price − Markdown
       62.66 = S − 35% of S
       62.66 = 0.65S
            S = $96.40

Total cost = Cost + Expense
       54.75 = C + 25% of S
       54.75 = C + 0.25(96.40)
       54.75 = C + 24.10
            C = $30.65

Profit = Revenue − Total cost
        = 62.66 − 54.75
        = $7.91

5.   Sale price = Regular selling price − Markdown
     120.00 = S − 25% of S
     120.00 = 0.75S
           S = $160.00

Profit = Revenue − Total cost
  −4.20 = 120.00 − TC

Total cost = $124.20

Total cost = Cost + Expense
   124.20 = 105.00 + E
       E = $19.20

Expense as a percent of Regular selling price

$$= \frac{19.20}{160.00}(100\%) = 12\%$$

CHAPTER 6

C. 1. (a) Cost $= (0.45)(0.75)(440.00) = \$148.50$

Regular selling price $=$ Cost $+$ Markup

$\qquad = C + 180\%$ of C

$\qquad = 2.8(148.50) = \$415.80$

Sale price $=$ Regular selling price $-$ Markdown

$\qquad = R - 45\%$ of R

$\qquad = (0.55)(415.80) = \$228.69$

(b) Markup realized $= 228.69 - 148.50 = \$80.19$

Rate of markup realized based on cost $= \dfrac{80.19}{148.50} = 0.54 \doteq 54\%$

3. Sale price $=$ Cost $+$ Markup

$\qquad S = C + 30\%$ of S

$\qquad S = 36.75 + 0.30S$

$\qquad 0.70S = 36.75$

$\qquad S = \$52.50$

Regular selling price $-$ Discount $= S$

$\qquad R - 25\%$ of $R = S$

$\qquad 0.75R = 52.50$

$\qquad R = \$70.00$

The regular selling price should be \$70.00

5. (a) Markdown $= 125.00 - 75.00 = \$50.00$

Rate of markdown $= \dfrac{50.00}{125.00} = 0.40 = 40\%$

(b) Cost $= (0.6\dot{6})(0.85)(120.00) = \$68.00$

Total cost $= C + E$

$\qquad = 68.00 + 12\%$ of 125.00

$\qquad = 68.00 + 15.00 = \$83.00$

Profit $=$ Revenue $-$ Total cost

$\qquad = 75.00 - 83.00 = \langle\$8.00\rangle$

(c) Rate of markup based on cost

$= \dfrac{75.00 - 68.00}{68.00} = 0.10294 = 10.3\%$

104

(d) Rate of markup based on sale price

$$= \frac{75.00 \ - \ 68.00}{75.00} \ = \ 0.93333 \ = \ 9\frac{1}{3}\%$$

7. Regular selling price $= \ C \ + \ E \ + \ P$

$$R \ = \ 36.40 \ + \ 24\% \text{ of } R \ + \ 20\% \text{ of } R$$
$$R \ = \ 36.40 \ + \ 0.44R$$
$$0.56R \ = \ 36.40$$
$$R \ = \ \$65.00$$

Sale price $= \ 65.00 \ - \ 30\% \text{ of } 65.00 \ = \ 0.70(65) \ = \ \$45.50$

Total cost $= \ 36.40 \ + \ 24\% \text{ of } 65.00 \ = \ 36.40 \ + \ 15.60 \ = \ \$52.00$

Profit $= \ 45.50 \ - \ 52.00 \ = \ \langle\$6.50\rangle$

9. (a) $\text{Cost} \ = \ (0.60)(0.83)(24.00) \ = \ \$12.00$

Regular selling price $= \ C \ + \ 25\% \text{ of } C \ + \ 33\frac{1}{3}\% \text{ of } C$

$$= \ 158\frac{1}{3}\%(12.00) \ = \ \$19.00$$

Regular selling price is \$19.00

(b) Break - even price
= Cost + Expense
= 12.00 + 25% of 12.00
= \$15.00
Maximum amount of markdown
= 19.00 - 15.00 = \$4.00

(c) Rate of markdown to break - even

$$= \frac{4.00}{19.00} \ = \ 0.21053 \ = \ 21.05\%$$

11. Cost $= (0.40)(0.83)(900.00) = \$300.00$

Regular selling price $= C + 0.45R + 0.15R$

$$0.40R = 300.00$$

$$R = \$750.00$$

New regular selling price $-$ Discount $= R$

$$N - 0.375N = 750.00$$

$$0.625N = 750.00$$

$$N = \$1200.00$$

Sale price $= N - 0.55N$

$$= 0.45(1200.00)$$

$$= \$540.00$$

Total cost $= 300.00 + 0.45R$

$$= 300.00 + 0.45(750.00)$$

$$= 300.00 + 337.50$$

$$= \$637.50$$

Profit $= 540.00 - 637.50$

$$= \langle \$97.50 \rangle$$

13. Cost $= (0.66)(0.95)(900.00) = \$570.00$

Regular selling price $= C + 0.15R + 0.09R$

$$R = 570.00 + 0.24R$$

$$0.76R = 570.00$$

$$R = \$750.00$$

New regular selling price $-$ Discount $= R$

$$N - 0.25N = 750.00$$

$$0.75N = 750.00$$

$$N = \$1000.00$$

Sale price $= N - 0.40N$

$$= 0.60(1000.00)$$

$$= \$600.00$$

Total cost $= 570.00 + 0.15(750.00)$

$$= 570.00 + 112.50$$

$$= 682.50$$

Profit $= 600.00 - 682.50$

$$= \langle \$82.50 \rangle$$

1. (a) Net price $= (0.75)(0.80)(0.95)(56.00) = \$31.92$

(b) Discount $= 56.00 - 31.92 = \$24.08$

(c) Single rate of discount $= \dfrac{24.08}{56.00} = 0.43 = 43\%$

3. Single rate
$= 1 - (0.65)(0.88)(0.95)$
$= 1 - 0.5434$
$= 0.4566$
$= 45.66\%$

5. Regular price $= (0.85)(112.00) = \$95.20$
Additional discount $= 95.20 - 80.92 = \$14.28$
Additional discount percent $= \dfrac{14.28}{95.20} = 0.15 = 15\%$

7. Net price $= (0.80)(0.85)L$
$\qquad 20.40 = (0.80)(0.85)L$
$\qquad\quad L = \$30.00$
List price is $30.00

9.

Invoice no.	Rate of discount		Net amount paid
312	—		$ 923.00
429	2%	(0.98)(784.00)	768.32
563	5%	(0.95)(873.00)	j 829.35
Remittance			$2520.67

11.   Net invoice = $(0.75)(0.80)(16\ 000.00) = \$9600.00$

(a)   Original balance $9600.00
       Payment Sept. 30   4600.00
                                           $5000.00
       Payment Oct. 20   3000.00
       Balance due $2000.00

(b)   Payment Sept. 30:
            allow 5%
       Paid = $(0.95)(4600.00)$ $4370.00
       Payment Oct. 20:
            allow 2%
       Paid = $(0.98)(3000.00)$ 2940.00
       Final payment   2000.00
            Total paid $9310.00

13.   (a)   Cost = $(0.80)(6.00) = \$4.80$
              $S = C + 0.45C + 0.20C$
              $S = 1.65C$
              $S = 1.65(4.80)$
              $S = \$7.92$
              Selling price is $7.92 per bag

       (b)   Amount of markup = $7.92 - 4.80 = \$3.12$

       (c)   Rate of markup based on selling price $= \dfrac{3.12}{7.92} = 0.3939 = 39.4\%$

       (d)   Rate of markup based on cost $= \dfrac{3.12}{4.80} = 0.65 = 65\%$

       (e)   Break-even price
              = Cost + Overhead
              = $4.80 + 0.45(4.80)$
              = $4.80 + 2.16$
              = $6.96

(f)   For a selling price of $6.00

$$\text{Profit} = 6.00 - 6.96$$
$$= \langle \$0.96 \rangle$$

15.   (a) $C + 0.35C = $ Selling price

$$1.35C = 8.91$$
$$C = 6.60$$

Cost was $6.60

(b)   Markup as a percent of selling price

$$= \frac{8.91 - 6.60}{8.91}(100\%) = 25.9\%$$

17.   (a) $\text{Cost} = (0.625)(0.82)(1800.00) = \$922.50$

$$C + 1.20C = \text{Regular selling price}$$
$$2.20C = R$$
$$R = 2.20(922.50)$$
$$R = \$2029.50$$

$$\text{Sale price} = 0.60R$$
$$= 0.60(2029.50)$$
$$= \$1217.70$$

(b)   Realized rate of markup based on cost

$$= \frac{1217.70 - 922.50}{922.50} = \frac{295.20}{922.50} = 0.32 = 32\%$$

19.   (a) $\text{Cost} = (0.60)(0.85)(36.00) = \$18.36$

$$C + 0.40S = S$$
$$18.36 = 0.60S$$
$$S = \$30.60$$

$$\text{Markdown} = 30.60 - 22.95 = \$7.65$$

$$\text{Rate of markdown} = \frac{7.65}{30.60} = 0.25 = 25\%$$

(b)   Total cost = C + 0.25(30.60)

$$= 18.36 + 7.65$$
$$= \$26.01$$

Profit = 22.95 − 26.01 = ⟨$3.06⟩

(c)   Rate of markup realized based on cost

$$= \frac{22.95 - 18.36}{18.36} = \frac{4.59}{18.36} = 0.25 = 25\%$$

21.   (a)   Cost = (0.60)(0.83)(0.92)(250.00) = \$115.00

R = C + 0.65C + 0.55C

$$= 2.20C = 2.20(115.00) = \$253.00$$

Regular selling price is \$253.00

(b)   Break - even price

$$= C + E$$
$$= 115.00 + 0.65(115.00)$$
$$= 115.00 + 74.75$$
$$= \$189.75$$

(c)   Maximum reduction = 253.00 − 189.75 = \$63.25

Maximum rate of markdown $= \dfrac{63.25}{253.00} = 0.25 = 25\%$

23.   Net price = (0.6$\dot{6}$)(0.85)(1860.00) = \$1054.00

Reduction required = 1054.00 − 922.25 = \$131.75

Additional discount $= \dfrac{131.75}{1054.00} = 0.125 = 12.5\%$

25.   (a)   37.5% of Cost = 42.00

$$0.375C = 42.00$$
$$C = \$112.00$$

Regular selling price = 112.00 + 42.00 = \$154.00

(b)   Rate of markup based on regular selling price

$$= \frac{42.00}{154.00} = 0.272727 = 27.27\%$$

(c)   Break - even price $= 112.00 + 17.5\%$ of 154.00
$$= 112.00 + 26.95$$
$$= \$138.95$$

(d)   Markdown $= 154.00 - 121.66 = \$32.34$
Rate of markdown $= \dfrac{32.34}{154.00} = 0.21 \doteq 21\%$

Self-Test

1.   Net price $= 0.625 \times 0.875 \times 0.9166667 \times 590.00 = \$295.77$

3.   Single rate $= 1.00 - (0.60)(0.90)(0.9166667)$
$$= 1.00 - 0.495 = 0.505 = 50.5\%$$

5.

Invoice	Net price		Discount	Pay
March 21	$(0.80)(0.90)(850.00)$	$= 612.00$	Nil	$ 612.00
April 10	$(0.70)(0.83)(960.00)$	$= 560.00$	2%	548.80
April 30	$(0.66)(0.75)(0.95)(1040.00)$	$= 494.00$	4%	474.24
			Remittance	$1635.04

7.   Discount allowed on partial payment: 4%
Payment $= 96\%$ of reduction in debt
$$1392.00 = 0.96R$$
$$R = 1450.00$$
Debt is reduced by $1450.00.

9. $\quad$ C + M = Regular selling price
$$C + 0.40R = R$$
$$180.00 = 0.60R$$
$$R = 300.00$$
$$\text{Sale price} = 0.80 \times 300.00$$
$$= \$240.00$$

11.
$$\text{Cost} = (0.65)(0.875)(350.00)$$
$$= 199.06$$
$$C + M = \text{Regular selling price}$$
$$199.06 + 1.5(199.06) = R$$
$$R = 497.65$$
$$\text{Sale price} = (0.70)(497.65) = \$348.36$$

13. $\quad$ Markup = 2520.00 − 900.00 = 1620.00
Rate of markup based on cost
$$= \frac{1620.00}{900.00} = 1.8 = 180\%$$

15. $\quad$ Markdown = 1560.00 − 1195.00 = 365.00
Rate of markdown
$$= \frac{365.00}{1560.00} = 0.2339744 = 23.40\%$$

17. $\quad$ Cost = (0.75)(0.85)(1480.00) = 943.50
$$C + 0.40R + 0.10R = R$$
$$943.50 + 0.50R = R$$
$$943.50 = 0.50R$$
$$R = 1887.00$$
Sale price = (0.55)(1887.00) = 1037.85
Total cost = 943.50 + (0.40)(1887.00) = 943.50 + 754.80 = 1698.30
Operating profit = 1037.85 − 1698.30 = −\$660.45(Loss)

7 Simple interest

Exercise 7.1

A. 1. $r = 0.035;\ t = 1.25$

    3. $r = 0.1025;\ t = \dfrac{165}{365}$

Exercise 7.2

A. 1. $14 + 28 + 31 + 30 + 9 = 112$ days

    3. $30 + 31 + 31 + 29 + 31 + 14 = 166$ days (leap year)

B. 1. April 1, 2001 → Day   91
        December 1, 2001 → Day   335
        Elapsed number   244

    3. April 5, 2003 → Day   95
        Number of days in 2003   365
        Days remaining   270
        March 11, 2004 → Day   71 (leap year)
        Elapsed number   341

Exercise 7.3

A. 1. $I = 4000.00 \times 0.105 \times 2.25 = \$945.00$

    3. $I = 1660.00 \times 0.0975 \times \dfrac{16}{12} = \$215.80$

    5. $I = 980.00 \times 0.115 \times \dfrac{244}{365} = \$75.34$

B. 1. Number of days = 1 + 31 + 31 + 28 + 31 + 30 + 4 = 156

$$I = 275.00 \times 0.0925 \times \frac{156}{365} = \$10.87$$

3. Number of days = 27 + 31 + 30 + 31 + 31 + 30 + 31 + 3 = 214

$$I = 424.23 \times 0.1275 \times \frac{214}{365} = \$31.71$$

Exercise 7.4

A. 1. $P = \dfrac{I}{RT} = \dfrac{67.83}{0.095 \times \frac{7}{12}} = \dfrac{67.83 \times 12}{0.095 \times 7} = \$1224.00$

3. $R = \dfrac{I}{PT} = \dfrac{215.00}{2400.00 \times \frac{10}{12}} = \dfrac{215.00 \times 12}{2400.00 \times 10} = 0.1075 = 10.75\%$

5. $T = \dfrac{I}{PR} = \dfrac{36.17}{954.00 \times 0.0325} = 1.1665860$ years

T(months) = 1.1665860 × 12 = 14

7. $T = \dfrac{I}{PR} = \dfrac{7.14}{344.75 \times 0.0525} = 0.3944888$ years

T(days) = 0.3944888 × 365 = 144

B. 1. $I = 148.32;\ r = 0.0675;\ t = \dfrac{8}{12}$

$$P = \frac{I}{rt} = \frac{148.32}{0.0675 \times \frac{8}{12}} = \frac{148.32 \times 12}{0.0675 \times 8} = \$3296.00$$

3. $P = 880.00;\ I = 104.50;\ t = \dfrac{15}{12}$

$R = \dfrac{I}{Pt} = \dfrac{104.50}{880.00 \times \frac{15}{12}} = \dfrac{104.50 \times 12}{880.00 \times 15} = 0.095 = 9.5\%$

5. $P = 1290.00;\ I = 100.51;\ r = 0.085$

$T(\text{months}) = \dfrac{I}{Pr} \times 12 = \dfrac{100.51 \times 12}{1290.00 \times 0.085} = 11$

7. Number of days $= 13 + 31 + 31 + 30 + 31 + 30 + 14 = 180;$

$I = 39.96;\ r = 0.0925;\ t = \dfrac{180}{365}$

$P = \dfrac{I}{rt} = \dfrac{39.96}{0.0925 \times \frac{180}{365}} = \dfrac{39.96 \times 365}{0.0925 \times 180} = \$876.00$

9. $I = 2000.00;\ r = 0.06;\ t = \dfrac{1}{12}$

$P = \dfrac{2000.00}{\frac{1}{12}(0.06)} = \$400\,000$

11. $P = 7800.00;\ I = 88.47;\ t = \dfrac{120}{365};$

$r = \dfrac{88.47}{7800.00\left(\frac{120}{365}\right)} = \dfrac{88.47(365)}{7800.00(120)} = 0.0344995 = 3.45\%$

13. $P = 3448.00;\ I = 3827.66 - (3448.00 + 344.80) = 34.86;\ r = 0.09;$

$t = \dfrac{34.86}{3448.00(0.09)} = 0.1123357$ years

Number of days overdue $= 0.1123357(365) = 41$

15. (a) $P = 24\,000.00;\ r = 0.035;$
time in days from April 1 to June 30 inclusive $= 30 + 31 + 30 = 91$

$I = 24\,000.00(0.035)\left(\dfrac{91}{365}\right) = \$209.42$

115

(b) P = Balance June 30 = 24 000.00 + 209.42 = 24 209.42; r = 0.035;
time in days from July 1 to September 30 inclusive = 31 + 31 + 30 = 92;

$$I = 24\,209.42(0.035)\left(\frac{92}{365}\right) = \$213.57$$

(c) The two reasons are:
(i) the principals are different
(ii) the time periods are different

Exercise 7.5

A. 1. P = 480.00; r = 0.035; $t = \dfrac{220}{365}$

$$S = P(1 + rt) = 480.00\left[1 + 0.035 \times \frac{220}{365}\right]$$
$$= 480.00(1 + 0.0210959)$$
$$= \$490.13$$

3. P = 820.00; r = 0.0475; $t = \dfrac{9}{12}$

$$S = P(1 + rt) = 820.00\left[1 + 0.0475 \times \frac{9}{12}\right]$$
$$= 820.00(1 + 0.035625)$$
$$= \$849.21$$

B. 1. P = 2500.00; r = 0.0345; $t = \dfrac{180}{365}$

$$S = P(1+rt) = 2500.00\left(1 + 0.0345 \times \frac{180}{365}\right)$$
$$= 2500.00(1 + 0.0170137)$$
$$= 2500.00(1.0170137)$$
$$= \$2542.53$$

3. S = 13 864.50; r = 0.0365; $t = \dfrac{270}{365}$

$$13\,864.50 = P\left(1 + 0.0365 \times \frac{270}{365}\right)$$
$$13\,864.50 = P(1 + 0.027)$$
$$13\,864.50 = 1.027P$$
$$P = \frac{13\,864.50}{1.027} = \$13\,500.00$$

Exercise 7.6

A. 1. $S = 279.30;\ r = 0.04;\ t = \dfrac{15}{12}$

$P = \dfrac{S}{1 + rt} = \dfrac{279.30}{1 + 0.04 \times \frac{15}{12}} = \dfrac{279.30}{1 + 0.05} = \dfrac{279.30}{1.05} = \$266.00$

$I = 279.30 - 266.00 = \$13.30$

3. $I = 29.67;\ r = 0.086;\ t = \dfrac{8}{12}$

$P = \dfrac{29.67}{0.086 \times \frac{8}{12}} = \dfrac{29.67 \times 12}{0.086 \times 8} = \$517.50$

$S = P + I = 517.50 + 29.67 = \$547.17$

5. $S = 2109.24;\ I = 84.24;\ r = 0.052$

$P = 2109.24 - 84.24 = \$2025.00$

$t(\text{days}) = \dfrac{84.24}{2025.00 \times 0.052} \times 365 = \dfrac{84.24 \times 365}{2025.00 \times 0.052} = 292$

B. 1. $S = 1241.86;\ r = 0.039;\ t = \dfrac{5}{12}$

$P = \dfrac{S}{1 + rt} = \dfrac{1241.86}{1 + 0.039 \times \frac{5}{12}} = \dfrac{1241.86}{1 + 0.01625} = \dfrac{1241.86}{1.01625} = \$1222.00$

3. $S = 1760.00;\ r = 0.0975;\ t = \dfrac{4}{12}$

$P = \dfrac{S}{1 + rt} = \dfrac{1760.00}{1 + 0.0975 \times \frac{4}{12}} = \dfrac{1760.00}{1 + 0.0325} = \dfrac{1760.00}{1.0325} = \$1704.60$

5. $S = 1750.00;\ r = 0.189;$

$t = 30 + 31 + 30 + 31 + 31 + 28 = 181\ (\text{days})$

$P = \dfrac{S}{1 + rt} = \dfrac{1750.00}{1 + 0.189 \times \frac{181}{365}} = \dfrac{1750.00}{1 + 0.0937233} = \$1600.04$

7. $S = 4242.00;\ r = 0.075;\ t = \dfrac{8}{12}$

$P = \dfrac{4242.00}{1 + 0.075 \times \frac{8}{12}} = \dfrac{4242.00}{1 + 0.05} = \$4040.00$

Amount saved $= 4242.00 - 4040.00 = \$202.00$

Exercise 7.7

A. 1. $P = 800.00;\ r = 0.11;\ t = \dfrac{4}{12};$ find S.

$S = P(1 + rt) = 800.00\left(1 + 0.11 \times \dfrac{4}{12}\right) = 800.00(1 + 0.0366667) = \$829.33$

3. $P = 600.00;\ r = 0.07;\ t = \dfrac{5}{12};$ find S.

$S = P(1 + rt) = 600.00\left(1 + 0.07 \times \dfrac{5}{12}\right) = 600.00(1 + 0.0291667) = \$617.50$

5. Let the size of the single replacement payment be x.
Focal date is today.
Dated value at the focal date of $500 due 4 months ago:

$P = 500.00;\ r = 0.12;\ t = \dfrac{4}{12};$ find S.

$S = P(1 + rt) = 500.00\left(1 + 0.12 \times \dfrac{4}{12}\right) = 500.00(1 + 0.04) = \$520.00$

Dated value at the focal date at of $600 due in 2 months:

$S = 600.00;\ r = 0.12;\ t = \dfrac{2}{12}$

$P = \dfrac{S}{1 + rt} = \dfrac{600.00}{1 + 0.12 \times \frac{2}{12}} = \dfrac{600.00}{1 + 0.02} = \$588.24$

Equivalent single replacement payment:
$x = 520.00 + 588.24 = \$1108.24$

7. Let the size of the payment due in 6 months be x.
   Focal date is today.
   The sum of the dated values of the original debts at the focal date
   = $2000.00.
   The sum of the dated values of the replacement payments at the focal date

$$= \frac{1200.00}{1 + 0.12 \times \frac{3}{12}} + \frac{x}{1 + 0.12 \times \frac{6}{12}}$$

$$= \frac{1200}{1.03} + \frac{x}{1.06}$$

$$= 1165.05 + \frac{x}{1.06}$$

Equation of equivalence:

$$2000.00 = 1165.05 + \frac{x}{1.06}$$

$$834.95 = \frac{x}{1.06}$$

$$x = 885.05$$

The balance due in 6 months is $885.05.

9. Let the size of the equal payments be x.
   Focal date is today.
   Dated value of the original debt at the focal date
   = 1200.00
   Sum of the dated values of the replacement payments at the focal date

$$= \frac{x}{1 + 0.10 \times \frac{3}{12}} + \frac{x}{1 + 0.10 \times \frac{6}{12}} = \frac{x}{1.025} + \frac{x}{1.05}$$

Equation of equivalence:

$$1200.00 = \frac{x}{1.025} + \frac{x}{1.05}$$

$$1200.00 = 0.9756098x + 0.9523810x$$

$$1200.00 = 1.9279908x$$

$$x = 622.41$$

The size of the equal payments is $622.41.

11. Let the size of the equal payments be x. .
The focal date is now.
The maturity value of \$1200 due in 8 months with 10% interest

$$= 1200.00\left(1 + 0.10 \times \frac{8}{12}\right) = 1280.00$$

The sum of the dated values of the original payments at the focal date

$$= 1500.00\left(1 + 0.12 \times \frac{4}{12}\right) + \frac{1280.00}{1 + 0.12 \times \frac{8}{12}}$$

$$= 1560.00 + 1185.19$$

The sum of the dated values of the replacement payments at the focal date

$$= 700.00 + \frac{x}{1 + 0.12 \times \frac{6}{12}} + \frac{x}{1 + 0.12 \times \frac{12}{12}}$$

$$= 700.00 + \frac{x}{1.06} + \frac{x}{1.12}$$

$$= 700.00 + 0.9433962x + 0.8928571x$$

The equation of equivalence is

$$1560.00 + 1185.19 = 700.00 + 0.9433962x + 0.8928571x$$
$$2745.19 = 700.00 + 1.8362533x$$
$$2045.19 = 1.8362533x$$
$$x = 1113.78$$

The size of the equal payments is \$1113.78.

B. 1. Let the size of the single payment be x.
The focal date is today.
The equation of equivalence is

$$x = \frac{600.00}{1 + 0.10 \times \frac{3}{12}} + \frac{600.00}{1 + 0.10 \times \frac{6}{12}}$$

$$x = 585.37 + 571.43$$
$$x = 1156.80$$

The single payment today is \$1156.80.

3. Let the size of the final payment be x.
The focal date is one month from now.
The equation of equivalence is

$$400.00\left(1 + 0.06 \times \frac{4}{12}\right) + 700.00\left(1 + 0.06 \times \frac{1}{12}\right)$$

$$= 600.00 + \frac{x}{1 + 0.06 \times \frac{3}{12}}$$

$$408.00 + 703.50 = 600.00 + \frac{x}{1.015}$$
$$1111.50 = 600.00 + 0.9852217x$$
$$511.50 = 0.9852217x$$
$$x = 519.17$$

The size of the final payment is \$519.17.

5. Let the size of the equal payments be x.
    The focal date is today.
    The equation of equivalence is

    $$4000.00 = \frac{x}{1 + 0.12 \times \frac{4}{12}} + \frac{x}{1 + 0.12 \times \frac{8}{12}} + \frac{x}{1 + 0.12 \times \frac{12}{12}}$$

    $$4000.00 = \frac{x}{1.04} + \frac{x}{1.08} + \frac{x}{1.12}$$

    $$4000.00 = 0.9615385x + 0.9259259x + 0.8928571x$$

    $$4000.00 = 2.7803215x$$

    $$x = 1438.68$$

    The size of the equal payments is $1438.68.

7. Let the size of the equal payments be x.
    Focal date is today.
    The maturity value of the second payment of $1000 due in 9 months with 9% interest

    $$= 1000.00\left(1 + 0.09 \times \frac{9}{12}\right) = 1067.50$$

    The equation of equivalence is

    $$800.00 + \frac{1067.50}{1 + 0.11 \times \frac{9}{12}} = \frac{x}{1 + 0.11 \times \frac{3}{12}} + \frac{x}{1 + 0.11 \times \frac{6}{12}}$$

    $$800.00 + 986.14 = \frac{x}{1.0275} + \frac{x}{1.055}$$

    $$1786.14 = 0.9732360x + 0.9478673x$$

    $$1786.14 = 1.9211033x$$

    $$x = 929.75$$

    The size of the equal payments is $929.75.

Review Exercise

1. (a) No. of days $= 6 + 31 + 30 + 31 + 31 + 30 + 13 = 172$

    (b) No. of days $= 2 + 31 + 30 + 31 + 30 + 31 + 31 + 0 = 186$

3. (a) $I = 34.80$; $r = 0.05$; $t = \frac{219}{365}$

    $$P = \frac{34.80}{0.05 \times \frac{219}{365}} = \frac{34.80 \times 365}{0.05 \times 219} = \$1160.00$$

    (b) No. of days $= 2 + 30 + 31 + 31 + 28 + 31 + 30 + 31 + 0 = 214$

    $I = 34.40$; $r = 0.0975$; $t = \frac{214}{365}$

    $$P = \frac{34.40}{0.0975 \times \frac{214}{365}} = \frac{34.40 \times 365}{0.0975 \times 214} = \$601.77$$

5. (a) $S = 665.60$; $r = 0.10$; $t = \dfrac{146}{365}$

$$P = \frac{S}{1 + rt} = \frac{665.60}{1 + 0.10 \times \frac{146}{365}} = \frac{665.60}{1 + 0.04} = \$640.00$$

(b) $S = 6300.00$; $r = 0.0775$; $t = \dfrac{16}{12}$

$$P = \frac{S}{1 + rt} = \frac{6300.00}{1 + 0.0775 \times \frac{16}{12}} = \frac{6300.00}{1 + 0.1033333} = \$5709.97$$

7. No. of days $= 30 + 31 + 30 + 31 + 31 + 28 + 31 + 29 = 241$

$I = 148.57$; $r = 0.075$; $t = \dfrac{241}{365}$

$$P = \frac{148.57}{0.075 \times \frac{241}{365}} = \frac{148.57 \times 365}{0.075 \times 241} = \$3000.17$$

9. No. of days $= 30 + 31 + 31 + 30 + 31 + 30 + 0 = 183$

$I = 1562.04 - 1500.00 = 62.04$; $P = 1500.00$; $t = \dfrac{183}{365}$

$$r = \frac{62.04}{1500.00 \times \frac{183}{365}} = \frac{62.04 \times 365}{1500.00 \times 183} = 0.0824940 = 8.25\%$$

11. $I = 3195.72 - 3100.00 = 95.72$; $P = 3100.00$; $r = 0.0575$

$$t(\text{days}) = \frac{95.72}{3100.00 \times 0.0575} \times 365 = 196$$

13. No. of days $= 21 + 31 + 31 + 30 + 31 + 30 + 14 = 188$

$P = 1550.00$; $r = 0.065$; $t = \dfrac{188}{365}$

$$S = 1550.00\left(1 + 0.065 \times \frac{188}{365}\right) = 1550.00(1 + 0.0334795) = \$1601.89$$

15. No. of days $= 30 + 31 + 31 + 29 + 31 + 30 + 30 = 212$

$S = 3367.28$; $r = 0.09$; $t = \dfrac{212}{365}$

$$P = \frac{S}{1 + rt} = \frac{3367.28}{1 + 0.09 \times \frac{212}{365}} = \frac{3367.28}{1 + 0.0522740} = \$3200.00$$

17. No. of days $= 30 + 31 + 31 + 30 + 14 = 136$

$S = 1785.00;\ r = 0.075;\ t = \dfrac{136}{365}$

$P = \dfrac{S}{1 + rt} = \dfrac{1785.00}{1 + 0.075 \times \frac{136}{365}} = \dfrac{1785.00}{1 + 0.0279452} = \$1736.47$

19. Let the size of the single payment be $\$x$.
The focal date is 30 days from now.

$$1450.00\left(1 + 0.09 \times \dfrac{75}{365}\right) + \dfrac{1200.00}{1 + 0.09 \times \frac{30}{365}} = x$$
$$1476.82 + 1191.19 = x$$
$$x = 2668.01$$

The size of the single payment is \$2668.01.

21. Let the size of the equal payments be $\$x$.
The focal date is today.

$$10\,000.00 = \dfrac{x}{1 + 0.075 \times \frac{60}{365}} + \dfrac{x}{1 + 0.075 \times \frac{120}{365}} + \dfrac{x}{1 + 0.075 \times \frac{180}{365}}$$
$$10\,000.00 = \dfrac{x}{1.0123288} + \dfrac{x}{1.0246575} + \dfrac{x}{1.0369863}$$
$$10\,000.00 = 0.9878213x + 0.9759359x + 0.9643329x$$
$$10\,000.00 = 2.9280901x$$
$$x = 3415.20$$

The size of the equal payments is \$3415.20.

23. Let the size of the equal payments be $\$x$.
The focal date is today.

$$\dfrac{5000.00}{1 + 0.095 \times \frac{12}{12}} = x + \dfrac{x}{1 + 0.095 \times \frac{6}{12}} + \dfrac{x}{1 + 0.095 \times \frac{12}{12}}$$
$$\dfrac{5000.00}{1.095} = x + \dfrac{x}{1.0475} + \dfrac{x}{1.095}$$
$$4566.21 = x + 0.9546539x + 0.9132420x$$
$$4566.21 = 2.8678959x$$
$$x = 1592.18$$

The size of the equal payments is \$1592.18.

25. Let the size of the single payment be $\$x$.
The focal date is 5 months from now.

$$700.00\left(1 + 0.09 \times \dfrac{8}{12}\right) + 1000.00\left(1 + 0.09 \times \dfrac{5}{12}\right) = 800.00\left(1 + 0.09 \times \dfrac{3}{12}\right) + x$$
$$742.00 + 1037.50 = 818.00 + x$$
$$x = 961.50$$

The size of the final payment is \$961.50.

27. Let the size of the equal payments be x.
    The focal date is today.
    The maturity value of $1500 due in 9 months with 7.5% interest

    $$= 1500.00\left(1 + 0.075 \times \frac{9}{12}\right) = 1584.38$$

    The maturity value of $1200 due in 18 months with 10.5% interest

    $$= 1200.00\left(1 + 0.105 \times \frac{18}{12}\right) = 1389.00$$

    $$2000.00\left(1 + 0.085 \times \frac{3}{12}\right) + \frac{1584.38}{1 + 0.085 \times \frac{9}{12}} + \frac{1389.00}{1 + 0.085 \times \frac{18}{12}}$$

    $$= x + \frac{x}{1 + 0.085 \times \frac{6}{12}} + \frac{x}{1 + 0.085 \times \frac{12}{12}}$$

    $$2042.50 + 1489.43 + 1231.93 = x + \frac{x}{1.0425} + \frac{x}{1.085}$$

    $$4763.86 = x + 0.9592326x + 0.9216590x$$

    $$4763.86 = 2.8808916x$$

    $$x = 1653.61$$

    The size of the equal payments is $1653.61.

Self-Test

1. $I = (1290.00)(0.035)\left(\dfrac{173}{365}\right) = \$21.40$

3. $r = \dfrac{81.25}{2500.00 \times \frac{6}{12}} = 0.065 = 6.5\%$

5. $S = 5500.00\left(1 + 0.0475 \times \dfrac{10}{12}\right) = 5500.00(1 + 0.0395833) = \$5717.71$

7. $PV = \dfrac{5000.00}{1 + 0.0725 \times \frac{243}{365}} = \dfrac{5000.00}{1 + 0.0482671} = \$4769.78$

9. $t = \dfrac{689.72}{8500.00 \times 0.0825} = 0.9835579 \text{ years} = 359 \text{ days}$

11. $P = \dfrac{7500.00}{1 + 0.0375 \times \frac{88}{365}} = \dfrac{7500.00}{1 + 0.0090411} = \$7432.80$

13. Let the size of the equal payments be x.

$$3320.00 = \frac{x}{1 + 0.0875 \times \frac{92}{365}} + \frac{x}{1 + 0.0875 \times \frac{235}{365}} + \frac{x}{1 + 0.0875 \times \frac{326}{365}}$$

$$3320.00 = \frac{x}{1.0220548} + \frac{x}{1.0563356} + \frac{x}{1.0781507}$$

$$3320.00 = 0.9784211x + 0.9466688x + 0.9275141x$$

$$3320.00 = 2.8526040x$$

$$x = 1163.85$$

15. Let the size of the final payment be x.

$$1010.00\left(1 + 0.0775 \times \frac{12}{12}\right) + 1280.00\left(1 + 0.0775 \times \frac{7}{12}\right)$$

$$= 615.00\left(1 + 0.0775 \times \frac{3}{12}\right) + x$$

$$1010.00(1.0775) + 1280.00(1.0452083) = 615.00(1.0193750) + x$$

$$1088.28 + 1337.87 = 626.92 + x$$

$$x = 1799.23$$

8   Simple interest applications

Exercise **8.1**

A.   1.   December 30, 2001

3. $530.00

5.   No. of days $= 2 + 31 + 28 + 31 + 30 + 31 + 1 = 154$

7. $S = 530.00 + 19.01 = \$549.01$

B.   1.   (a)   March 3, 2000

(b)   No. of days $= 1 + 31 + 30 + 31 + 31 + 29 + 2 = 155$

(c) $I = 840.00 \times 0.09 \times \dfrac{155}{365} = \$32.10$

(d)   Maturity value $= 840.00 + 32.10 = \$872.10$

3.   (a)   April 3, 2000

(b)   No. of days $= 63$

(c) $I = 1250.00 \times 0.105 \times \dfrac{63}{365} = \$22.65$

(d)   Maturity value $= 1250.00 + 22.65 = \$1272.65$

Exercise 8.2

A. 1. Due date is September 28.
No. of days = 7 + 30 + 31 + 31 + 27 = 126

$P = 620.00; \; r = 0.115; \; t = \dfrac{126}{365}$

$S = 620.00\left(1 + 0.115 \times \dfrac{126}{365}\right) = 620.00(1 + 0.0396986) = \$644.61$

3. No. of days = 153

$P = 820.00; \; r = 0.05; \; t = \dfrac{153}{365}$

$S = 820.00\left(1 + 0.05 \times \dfrac{153}{365}\right) = 820.00(1 + 0.0209589) = \$837.19$

Exercise 8.3

A. 1. Due date is October 12.
No. of days = 22 + 31 + 30 + 31 + 31 + 30 + 11 = 186

$S = 475.87; \; r = 0.09; \; t = \dfrac{186}{365}$

$P = \dfrac{475.87}{1 + 0.09 \times \frac{186}{365}} = \dfrac{475.87}{1 + 0.0458630} = \$455.00$

B. 1. Due date is November 13
No. of days = 22 + 30 + 31 + 12 = 95

$S = 1200.00; \; r = 0.11; \; t = \dfrac{95}{365}$

$P = \dfrac{1200.00}{1 + 0.11 \times \frac{95}{365}} = \dfrac{1200.00}{1 + 0.0286301} = \$1166.60$

3. Maturity value:
Due date is December 30.
Interest period is October 28–December 30, i.e., 63 days.

$P = 1600.00; \; r = 0.11; \; t = \dfrac{63}{365}$

$S = 1600.00\left(1 + 0.11 \times \dfrac{63}{365}\right) = 1600.00(1 + 0.0189863) = \$1630.38$

Present value on November 30:

$S = 1630.38; \; r = 0.09; \; t = \dfrac{30}{365}$

$P = \dfrac{1630.38}{1 + 0.09 \times \frac{30}{365}} = \dfrac{1630.38}{1 + 0.0073973} = \$1618.41$

CHAPTER 8

Exercise 8.4

A. 1. Due date is August 2.
Discount period is June 10–August 2, i.e., 53 days.

$$S = 1000.00; \; r = 0.075; \; t = \frac{53}{365}$$

$$P = \frac{1000.00}{1 + 0.075 \times \frac{53}{365}} = \frac{1000.00}{1 + 0.0108904} = \$989.23$$

Proceeds = \$989.23
Discount = 1000.00 − 989.23 = \$10.77

3. Maturity value:
Due date is September 4.
Interest period is June 1–September 4, i.e., 30 + 31 + 31 + 3 = 95 days.

$$P = 850.00; \; r = 0.09; \; t = \frac{95}{365}$$

$$S = 850.00\left(1 + 0.09 \times \frac{95}{365}\right) = 850.00(1 + 0.0234247) = \$869.91$$

Proceeds:
Discount period is July 20–September 4, i.e., 12 + 31 + 3 = 46 days.

$$S = 869.91; \; r = 0.115; \; t = \frac{46}{365}$$

$$P = \frac{869.91}{1 + 0.115 \times \frac{46}{365}} = \frac{869.91}{1 + 0.0144932} = \$857.48$$

Proceeds = \$857.48
Discount = 869.91 − 857.48 = \$12.43

Exercise 8.5

A. 1. $S = 100\,000.00; \; r = 0.0316; \; t = \frac{182}{365}$

$$P = \frac{100\,000.00}{1 + 0.0316\left(\frac{182}{365}\right)} = \frac{100\,000.00}{1.0157567} = \$98\,448.77$$

The price of the T-Bill is \$98 448.77.

3. $S = 5000.00; \; P = 4913.45; \; t = \frac{91}{365}$

Amount of yield = 5000.00 − 4913.45 = \$86.55

$$\text{Yield rate} = \frac{86.55}{4913.45\left(\frac{91}{365}\right)} = \frac{86.55(365)}{4913.45(91)} = 0.0706532 = 7.065\%$$

5.　(a)　$S = 100\,000.00; \ P = 99\,024.56; \ t = \dfrac{91}{365}$

Amount of yield $= 100\,000.00 - 99\,024.56 = \$975.44$

Yield rate $= \dfrac{975.44}{99\,024.56\left(\frac{91}{365}\right)} = \dfrac{975.44(365)}{99\,024.56(91)} = 0.0395102 = 3.95\%$

The original yield on the T-Bill is 3.95%.

(b)　Time to maturity at the date of sale is $91 - 42 = 49$ days.

$S = 100\,000.00; \ r = 0.03725; \ t = \dfrac{49}{365}$

$P = \dfrac{100\,000.00}{1 + 0.03725\left(\frac{49}{365}\right)} = \dfrac{100\,000.00}{1.0050007} = \$99\,502.42$

The price of the T-Bill is \$99 502.42.

(c)　The investment grew from \$99 024.56 to \$99 502.42 in 42 days.

$S = 99\,502.42; \ P = 99\,024.56; \ t = \dfrac{42}{365}$

The amount of yield $= 99\,502.42 - 99\,024.56 = \$477.86$

Yield rate $= \dfrac{477.86}{99\,024.56\left(\frac{42}{365}\right)} = \dfrac{477.86(365)}{99\,024.56(42)} = 0.0419374 = 4.194\%$

The rate of return realized on the T-Bill is 4.194%.

Exercise 8.6

A.　1.　Monthly payment $= 1500.00 \div 5 = \$300.00$
Monthly rate of interest $= 0.09 \div 12 = 0.0075$

Month	Loan amount owing	Interest for month
1	$1500.00	$1500.00 \times 0.0075 = \$11.25$
2	$1500.00 - 300.00 = 1200.00$	$1200.00 \times 0.0075 = \ \ 9.00$
3	$1200.00 - 300.00 = \ \ 900.00$	$900.00 \times 0.0075 = \ \ 6.75$
4	$900.00 - 300.00 = \ \ 600.00$	$600.00 \times 0.0075 = \ \ 4.50$
5	$600.00 - 300.00 = \ \ 300.00$	$300.00 \times 0.0075 = \ \ 2.25$
		Total interest　$33.75

3.

Date	Interest period	Principal	Rate	Interest	
June 10	May 10–June 10	8000.00	0.08	$8000.00 \times 0.08 \times \dfrac{31}{365} =$	$ 54.36
July 10	June 10–July 10	8000.00	0.08	$8000.00 \times 0.08 \times \dfrac{30}{365} =$	$ 52.60
Aug. 10	July 10–July 20	8000.00	0.08	$8000.00 \times 0.08 \times \dfrac{10}{365} =$	$ 17.53
	July 20–July 31 incl.	6000.00	0.08	$6000.00 \times 0.08 \times \dfrac{12}{365} =$	$ 15.78
	Aug. 1–Aug. 10	6000.00	0.095	$6000.00 \times 0.095 \times \dfrac{9}{365} =$	$ 14.05
					$ 47.36
Sept. 10	Aug. 10–Sept. 10	6000.00	0.095	$6000.00 \times 0.095 \times \dfrac{31}{365} =$	$ 48.41
Oct. 10	Sept. 10–Sept. 30 incl.	6000.00	0.095	$6000.00 \times 0.095 \times \dfrac{21}{365} =$	$ 32.79
	Oct. 1–Oct. 10	3000.00	0.085	$3000.00 \times 0.085 \times \dfrac{9}{365} =$	$ 6.29
					$ 39.08
Nov. 10	Oct. 10–Nov. 10	3000.00	0.085	$3000.00 \times 0.085 \times \dfrac{31}{365} =$	$ 21.66
Dec. 1	Nov. 10–Dec. 1	3000.00	0.085	$3000.00 \times 0.085 \times \dfrac{21}{365} =$	$ 14.67
				Total interest	$278.14

B. 1.

March 10	Original balance			$6000.00
June 30	Partial payment		$2000.00	
	Less interest due			
	$6000.00 \times 0.11 \times \dfrac{112}{365}$		202.52	1797.48
	Balance			$4202.52
Sept. 5	Partial payment		$2500.00	
	Less interest due			
	$4202.52 \times 0.11 \times \dfrac{67}{365}$		84.86	2415.14
	Balance			$1787.38
Nov. 15	Interest due			
	$1787.38 \times 0.11 \times \dfrac{71}{365}$			38.25
	Final payment			$1825.63

3.
March 25	Original balance		$20 000.00
May 15	Partial payment	$600.00	
	Less interest		

$$20\,000.00 \times 0.07 \times \frac{51}{365}$$

		195.62	404.38
	Balance		$19 595.62
June 30	Partial payment	$800.00	
	Less interest		

$$19\,595.62 \times 0.07 \times \frac{46}{365}$$

		172.87	627.13
	Balance		$18 968.49
Oct. 10	Partial payment	$400.00	
	Less interest		

$$18968.49 \times 0.07 \times \frac{1}{365} = \$\ 3.64$$

$$18\,968.49 \times 0.085 \times \frac{62}{365} = 273.87$$

$$18\,968.49 \times 0.095 \times \frac{39}{365} = 192.54$$

			470.05
	Unpaid interest		$ 70.05

$$\text{Oct. 31 incl. } 18\,968.49 \times 0.095 \times \frac{22}{365}$$

		108.61
Payment		$178.66

Exercise 8.7

A. 1. (a) Interest earned (on positive balances)
March 10 to March 15 inclusive: 6 days at 3.5% on $572.29

$$I = 572.29(0.035)\left(\frac{6}{365}\right) = \$0.33$$

March 16 to March 19 inclusive: 4 days at 3.5% on $307.29

$$I = 307.29(0.035)\left(\frac{4}{365}\right) = \$0.12$$

Total interest earned = 0.33 + 0.12 = $0.45

(b) Line of credit interest charged (on negative balances up to $1000.00)
March 1: 1 day at 11% on $527.71

$$I = 527.71(0.11)\left(\frac{1}{365}\right) = \$0.16$$

March 2 to March 9 inclusive: 8 days at 11% on $1000.00

$$I = 1000.00(0.11)\left(\frac{8}{365}\right) = \$2.41$$

March 20 to March 21 inclusive: 2 days at 11% on $692.71

$$I = 692.71(0.11)\left(\frac{2}{365}\right) = \$0.42$$

March 22 to March 26 inclusive: 5 days at 11% on $776.21

$$I = 776.21(0.11)\left(\frac{5}{365}\right) = \$1.17$$

March 27 to March 31 inclusive: 5 days at 11% on $941.21

$$I = 941.21(0.11)\left(\frac{5}{365}\right) = \$1.42$$

Total line of credit interest charged
$$= 0.16 + 2.41 + 0.42 + 1.17 + 1.42 = \$5.58$$

(c) Overdraft interest charged on negative balance in excess of $1000.00
March 2 to March 4 inclusive: 3 days at 24% on $127.71

$$\text{Overdraft interest} = 127.71(0.24)\left(\frac{3}{365}\right) = \$0.25$$

March 5 to March 9 inclusive: 5 days at 24% on $427.71

$$\text{Overdraft interest} = 427.71(0.24)\left(\frac{5}{365}\right) = \$1.41$$

Total overdraft interest $= 0.25 + 1.41 = \$1.66$

(d) Two transactions causing an overdraft or an overdraft to continue
Service charge $= 2(5.00) = \$10.00$

(e) The account balance on March 31
$$= -941.21 + 0.45 - 5.58 - 1.66 - 10.00 = -\$958.00$$

Exercise 8.8

A. 1.

Payment number	Balance before payment	Amount paid	Interest paid 0.708333%	Principal repaid	Balance after payment
0					1200.00
1	1200.00	180.00	8.50	171.50	1028.50
2	1028.50	180.00	7.29	172.71	855.79
3	855.79	180.00	6.06	173.94	681.85
4	681.85	180.00	4.83	175.17	506.68
5	506.68	180.00	3.59	176.41	330.27
6	330.27	180.00	2.34	177.66	152.61
7	152.61	153.69	1.08	152.61	—
Totals		1233.69	33.69	1200.00	

Exercise 8.9

A. 1. Number of outstanding payments $= 12 - 5 = 7$;
    the sum-of-the-digits of the 7 outstanding payments
    $$= \frac{7(7 + 1)}{2} = \frac{56}{2} = 28;$$
    the sum-of-the-digits for a 12-month lease
    $$= \frac{12(12 + 1)}{2} = \frac{156}{2} = 78;$$
    the rebate as a fraction of the total financing cost
    $$= \frac{28}{78}$$

  3. Number of outstanding payments $= 60 - 30 = 30$;
    the sum-of-the-digits of the 30 outstanding payments
    $$= \frac{30(30 + 1)}{2} = \frac{930}{2} = 465;$$
    the sum-of-the-digits for a 60-month lease
    $$= \frac{60(60 + 1)}{2} = \frac{3660}{2} = 1830;$$
    the rebate as a fraction of the total financing cost
    $$= \frac{465}{1830}$$

B. 1. Number of outstanding payments $= 18 - 11 = 7$;
    the sum-of-the-digits of the 7 outstanding payments
    $$= \frac{7(7 + 1)}{2} = \frac{56}{2} = 28;$$
    the sum-of-the-digits for an 18-month lease
    $$= \frac{18(18 + 1)}{2} = \frac{342}{2} = 171;$$
    the rebate as a fraction of the total financing cost
    $$= \frac{465}{1830}$$
    Total amount to be paid $\quad = \quad 720(18) \quad = \quad 12\,960.00$
    Total amount financed $\qquad\qquad = \quad \underline{12\,000.00}$
    $\qquad$ Finance charge $\qquad\qquad\qquad 960.00$

    Rebate $= \dfrac{28}{171}(960.00) = 157.19$

    Payout figure $= 7(960.00) - 157.19 = \$6562.81$.

Review Exercise

1. (a) Due date: November 2

   (b) Interest period: June 30–November 2
   $$1 + 31 + 31 + 30 + 31 + 1 = 125 \text{ days}$$
   $$I = 1600.00 \times 0.065 \times \frac{125}{365} = \$35.62$$

   (c) Maturity value = $1600.00 + 35.62 = \$1635.62$

3. $S = 767.68;\ r = 0.0925;\ t = \dfrac{93}{365}$
   $$P = \frac{S}{1 + rt} = \frac{767.68}{1 + 0.0925 \times \frac{93}{365}} = \frac{767.68}{1 + 0.0235685} = \$750.00$$

5. $P = \$5000.00;\ r = 0.06;\ t = \dfrac{153}{365}$
   $$S = P(1 + rt) = 5000.00\left(1 + 0.06 \times \frac{153}{365}\right)$$
   $$= 5000.00(1 + 0.0251507)$$
   $$= 5000.00(1.0251507)$$
   $$= \$5125.75$$

7. Maturity value
   Due date: November 1
   Interest period: July 31–November 1: 93 days
   $$P = 800.00;\ r = 0.08;\ t = \frac{93}{365}$$
   $$S = P(1 + rt) = 800.00\left(1 + 0.08 \times \frac{93}{365}\right) = 800.00(1 + 0.0203836) = \$816.31$$
   Present value:
   Discount period: October 20–November 1: 12 days
   $$S = 816.31;\ r = 0.08;\ t = \frac{12}{365}$$
   $$P = \frac{S}{1 + rt} = \frac{816.31}{1 + 0.08 \times \frac{12}{365}} = \frac{816.31}{1 + 0.0026301} = \$814.17$$

9. Due date: July 13
   Interest period: March 10–July 13
   $$22 + 30 + 31 + 30 + 12 = 125 \text{ days}$$

   $$S = 1300.00; \ r = 0.095; \ t = \frac{125}{365}$$

   $$P = \frac{S}{1 + rt} = \frac{1300.00}{1 + 0.095 \times \frac{125}{365}} = \frac{1300.00}{1 + 0.0325342} = \$1259.04$$

11. $$S = 25\,000.00; \ P = 24\,256.25; \ t = \frac{182}{365}$$

   Amount of yield $= 25\,000.00 - 24\,256.25 = \$743.75$

   Rate of return $= \dfrac{743.75}{24\,256.25\left(\frac{182}{365}\right)} = \dfrac{743.75(365)}{24\,256.25(182)} = 0.0614929 = 6.1493\%$

13.

Date	Interest period	Principal	Rate	Interest		
Mar. 31	Mar. 10–Mar. 31 incl.	10 000.00	9%	$10\,000.00 \times 0.09 \times \dfrac{22}{365}$	=	$ 54.25
Apr. 30	Apr. 1–Apr. 30 incl.	10 000.00	9%	$10\,000.00 \times 0.09 \times \dfrac{30}{365}$	=	$ 73.97
May 31	May 1–May 31 incl.	10 000.00	9%	$10\,000.00 \times 0.09 \times \dfrac{31}{365}$	=	$ 76.44
June 30	June 1–June 19 incl.	10 000.00	10%	$10\,000.00 \times 0.10 \times \dfrac{19}{365}$	=	$ 52.05
	June 20–June 30 incl.	7 000.00	10%	$7\,000.00 \times 0.10 \times \dfrac{11}{365}$	=	$ 21.10
						$ 73.15
July 31	July 1–July 31 incl.	7 000.00	10%	$7\,000.00 \times 0.10 \times \dfrac{31}{365}$	=	$ 59.45
Aug. 31	Aug. 1–Aug. 31 incl.	7 000.00	10%	$7\,000.00 \times 0.10 \times \dfrac{31}{365}$	=	$ 59.45
Sept. 30	Sept. 1–Sept. 30 incl.	3 000.00	10%	$3\,000.00 \times 0.10 \times \dfrac{30}{365}$	=	$ 24.66
Oct. 31	Oct. 1–Oct. 31 incl.	3 000.00	8%	$3\,000.00 \times 0.08 \times \dfrac{31}{365}$	=	$ 20.38
Nov. 15	Nov. 1–Nov. 15	3 000.00	8%	$3\,000.00 \times 0.08 \times \dfrac{14}{365}$	=	$ 9.21
					Total interest	$450.96

15.   (a)   Balance July 1   –$8195.00

Balance July 15   –8195.00 + 300.00 = –$7895.00

Interest charged July 31:

$$8195.00(0.10)\left(\frac{14}{365}\right) = \$31.43$$

$$7895.00(0.10)\left(\frac{17}{365}\right) = \underline{\ \ 36.77}$$

$$\$68.20$$

Balance July 31   –7895.00 – 68.20 = –$7963.20

Balance August 15   –7963.20 + 300.00 = –$7663.20

Balance August 20   –7663.20 – 3000.00 = –$10 663.20

Interest charged August 31:

$$7963.20(0.10)\left(\frac{14}{365}\right) = \$30.54$$

$$7663.20(0.10)\left(\frac{5}{365}\right) = \ \ 10.50$$

$$10\,000.00(0.10)\left(\frac{12}{365}\right) = \ \ 32.88$$

$$663.20(0.24)\left(\frac{12}{365}\right) = \underline{\ \ \ 5.23}$$

$$\$79.15$$

Balance August 31   –10 663.20 – 79.15 = –$10 742.35

Balance September 15   –10 742.35 + 300.00 = –$10 442.35

Interest charged September 30:

$$10\,000.00(0.10)\left(\frac{14}{365}\right) = \$38.36$$

$$10\,000.00(0.095)\left(\frac{16}{365}\right) = \ \ 41.64$$

$$742.35(0.24)\left(\frac{14}{365}\right) = \ \ \ 6.83$$

$$442.35(0.24)\left(\frac{16}{365}\right) = \underline{\ \ \ 4.65}$$

$$\$91.48$$

Balance September 30   –10 442.35 – 91.48 = –$10 533.83

Balance October 15   –10 533.83 + 300.00 = –$10 233.83

Balance October 25   –10 233.83 – 600.00 = –$10 833.83

Interest charged October 31:

$$10\,000.00(0.095)\left(\frac{31}{365}\right) = \$80.68$$

$$533.83(0.24)\left(\frac{14}{365}\right) = \ \ 4.91$$

$$233.83(0.24)\left(\frac{10}{365}\right) = \ \ 1.54$$

$$833.83(0.24)\left(\frac{7}{365}\right) = \underline{\ \ 3.84}$$

$$\$90.97$$

Balance October 31 $-10\,833.83 - 90.97 = -\$10\,924.80$
Balance November 15 $-10\,924.80 + 300.00 = -\$10\,624.80$
Interest charged November 30:

$$10\,000.00(0.095)\left(\frac{30}{365}\right) = \$78.08$$

$$924.80(0.24)\left(\frac{14}{365}\right) = 8.51$$

$$624.80(0.24)\left(\frac{16}{365}\right) = \underline{6.57}$$

$$\$93.16$$

(b) Balance November 30 $-10\,624.80 - 93.16 = -\$10\,717.96$

Self-Test

1. Due date: June 13, 2004
    Interest period: January 10–June 13
    $22 + 29 + 31 + 30 + 31 + 12 = 155$ days
    $$I = 565.00 \times 0.0825 \times \frac{155}{365} = \$19.79$$

3. Due date: September 23, 2003
    Interest period: May 20–September 23
    $12 + 30 + 31 + 31 + 22 = 126$ days
    $$P = \frac{1190.03}{1 + 0.075 \times \frac{126}{365}} = \frac{1190.03}{1 + 0.0258904} = \$1160.00$$

5. Due date: September 14, 2002 + 183 days
    $17 + 31 + 30 + 31 + 31 + 28 + 15$, i.e., March 16, 2003
    SONDJFM
    $$\text{Maturity value} = 1665.00\left(1 + 0.0925 \times \frac{183}{365}\right)$$
    $$= 1665.00(1 + 0.0463767) = \$1742.22$$
    Discount period: October 18, 2002–March 16, 2003
    $14 + 30 + 31 + 31 + 28 + 15 = 149$ days
    $$PV = \frac{1742.22}{1 + 0.105 \times \frac{149}{365}} = \frac{1742.22}{1 + 0.0428630} = \$1670.61$$

7.   (a) $S = 100\,000.00;\ r = 0.0385;\ t = \dfrac{182}{365}$

$$P = \frac{100\,000.00}{1 + 0.0385\left(\frac{182}{365}\right)} = \frac{100\,000.00}{1 + 0.0191973} = \$98\,116.43$$

    (b)   Time to maturity $= 182 - 67 = 115$ days

$S = 100\,000.00;\ P = 98\,853.84;\ t = \dfrac{115}{365}$

Amount of yield $= 100\,000.00 - 98\,853.84 = \$1146.16$

Yield rate $= \dfrac{1146.16}{98\,853.84\left(\frac{115}{365}\right)} = \dfrac{1146.16(365)}{98\,853.84(115)} = 0.0367999 = 3.680\%$

9.

March 1	Original balance			\$24 000.00
Apr. 15	Partial payment		\$600.00	
	Less interest			
	$24\,000.00 \times 0.07 \times \dfrac{45}{365}$		207.12	392.88
	Balance			\$23 607.12
July 20	Partial payment		\$400.00	
	Less interest			
	$23\,607.12 \times 0.07 \times \dfrac{96}{365}$		434.63	
	Unpaid interest		\$ 34.63	
Oct. 10	Partial payment		\$400.00	
	Less interest			
	July 20–July 31:			
	$23\,607.12 \times 0.07 \times \dfrac{12}{365}$	$= \$ 54.33$		
	Aug. 1–Oct. 10:			
	$23\,607.12 \times 0.085 \times \dfrac{70}{365}$	$= 384.83$		
	Unpaid interest, July 20	34.63	473.79	
	Unpaid interest		\$ 73.79	
Nov. 30	Interest			
	Oct. 10–Oct. 31:			
	$23\,607.12 \times 0.085 \times \dfrac{22}{365}$	$= \$120.95$		
	Nov. 1–Nov. 30:			
	$23\,607.12 \times 0.075 \times \dfrac{30}{365}$	$= 145.52$		
	Unpaid interest, Oct. 10	73.79		
	Payment due	\$340.26		

11.

Payment number	Balance before payment	Amount paid	Interest paid 0.5416667%	Principal repaid	Balance after payment
0					4000.00
1	4000.00	750.00	21.67	728.33	3271.67
2	3271.67	750.00	17.72	732.28	2539.39
3	2539.39	750.00	13.76	736.24	1803.15
4	1803.15	750.00	9.77	740.23	1062.92
5	1062.92	750.00	5.76	744.24	318.68
6	318.68	320.41	1.73	318.68	—
Totals		4070.41	70.41	4000.00	

PART THREE   *Mathematics of finance and investment*

9   Compound interest—Future value and present value

Exercise 9.1

A. 1. $m = 1;\ i = 12\% = 0.12; n = 5$

3. $m = 4;\ i = \dfrac{5.5\%}{4} = 1.375\% = 0.01375;\ n = 9 \times 4 = 36$

5. $m = 2;\ i = \dfrac{11.5\%}{2} = 5.75\% = 0.0575;\ n = 13.5 \times 2 = 27$

7. $m = 12;\ i = \dfrac{8\%}{12} = 0.6666667\% = 0.0066667;\ n = 12.5 \times 12 = 150$

9. $m = 2;\ i = \dfrac{12.25\%}{2} = 6.125\% = 0.06125;\ n = \dfrac{54}{12} \times 2 = 9$

B. 1. $(1 + 0.12)^5 = 1.12^5 = 1.7623417$

3. $(1 + 0.01375)^{36} = 1.01375^{36} = 1.6349754$

5. $(1 + 0.0575)^{27} = 1.0575^{27} = 4.5244954$

7. $(1 + 0.0066667)^{150} = 1.0066667^{150} = 2.7092894$

9. $(1 + 0.06125)^9 = 1.06125^9 = 1.7074946$

C. 1. (a) $n = 12 \times 4 = 48$

    (b) $i = \dfrac{10\%}{4} = 2.5\%$

    (c) 1.025^{48}

    (d) $1.025^{48} = 3.2714896$

Exercise 9.2

A. 1. $P = 400.00;\ m = 1;\ i = 7.5\% = 0.075;\ n = 8$
       $S = P(1 + i)^n = 400.00(1.075)^8 = 400.00(1.7834778) = \$713.39$

    3. $P = 1250.00;\ m = 4;\ i = \dfrac{6.5\%}{4} = 1.625\% = 0.01625;\ n = 9 \times 4 = 36$
       $S = 1250.00(1.01625)^{36} = 1250.00(1.7865703) = \$2233.21$

    5. $P = 1700.00;\ m = 4;\ i = \dfrac{8\%}{4} = 2.0\% = 0.02;\ n = 14.75 \times 4 = 59$
       $S = 1700.00(1.02)^{59} = 1700.00(3.2166969) = \$5468.38$

    7. $P = 2500.00;\ m = 12;\ i = \dfrac{8\%}{12} = 0.6666667\% = 0.0066667;\ n = 12.25 \times 12 = 147$
       $S = 2500.00(1.0066667)^{147} = 2500.00(2.6558178) = \$6639.54$

    9. $P = 480.00;\ m = 2;\ i = \dfrac{9.4\%}{2} = 4.7\% = 0.047;\ n = \dfrac{42}{12} \times 2 = 7$
       $S = 480.00(1.047)^7 = 480.00(1.3791985) = \$662.02$

B. 1. $P = 5000.00$; $m = 2$; $i = \dfrac{6.5\%}{2} = 3.25\% = 0.0325$; $n = 5 \times 2 = 10$

$S = 5000.00(1.0325)^{10} = 5000.00(1.3768943) = \$6884.47$

$I = 6884.47 - 5000.00 = \$1884.47$

3. $P = 1000.00$; $m = 12$; $i = \dfrac{5.5\%}{12} = 0.4583333\% = 0.0045833$

December 1, 1996 – August 1, 2003 contains 6 years, 8 months.

$n = 6 \times 12 + 8 = 80$

$S = 1000.00(1.0045833)^{80} = 1000.00(1.4417048) = \$1441.70$

5. $P = 100.00$

(a) $m = 1$; $i = 9\% = 0.09$; $n = 8$

$S = 100.00(1.09)^{8} = 100.00(1.9925626) = \$199.26$

(b) $m = 2$; $i = \dfrac{9\%}{2} = 4.5\% = 0.045$; $n = 8 \times 2 = 16$

$S = 100.00(1.045)^{16} = 100.00(2.0223702) = \$202.24$

(c) $m = 4$; $i = \dfrac{9\%}{4} = 2.25\% = 0.0225$; $n = 8 \times 4 = 32$

$S = 100.00(1.0225)^{32} = 100.00(2.0381030) = \$203.81$

(d) $m = 12$; $i = \dfrac{9\%}{12} = 0.75\% = 0.0075$; $n = 8 \times 12 = 96$

$S = 100.00(1.0075)^{96} = 100.00(2.0489212) = \$204.89$

7. $P = 100.00$; $m = 4$; $i = \dfrac{8\%}{4} = 2\% = 0.02$

   (a) $n = 5 \times 4 = 20$
   $S = 100.00(1.02)^{20} = 100.00(1.4859474) = \$148.59$
   Interest $= 148.59 - 100.00 = \$48.59$

   (b) $n = 10 \times 4 = 40$
   $S = 100.00(1.02)^{40} = 100.00(2.2080397) = \$220.80$
   Interest $= 220.80 - 100.00 = \$120.80$

   (c) $n = 20 \times 4 = 80$
   $S = 100.00(1.02)^{80} = 100.00(4.8754392) = \$487.54$
   Interest $= 487.54 - 100.00 = \$387.54$

9. $P = 98.5$; $m = 1$; $i = 3\% = 0.03$; $n = 10$
   $S = 98.5(1.03)^{10} = 98.5(1.3439164) = 132.38$
   Index at the beginning of 2001 would be 132.4.

11. (a) Bank: $P = 5000.00$; $m = 2$; $i = \dfrac{6.75\%}{2} = 3.375\% = 0.03375$; $n = 5 \times 2 = 10$
   $S = 5000.00(1.03375)^{10} = 5000.00(1.3936548) = \$6968.27$

   CU: $P = 5000.00$; $m = 12$; $i = \dfrac{6.5\%}{12} = 0.5416667\% = 0.0054167$;
   $n = 5 \times 12 = 60$
   $S = 5000.00(1.0054167)^{60} = 5000.00(1.3828201) = \$6914.10$
   The bank investment gives more interest.

   (b) Difference in interest $= 6968.27 - 6914.10 = \$54.17$

C. 1. Balance after 2.5 years:
   $P = 2000.00$; $m = 4$; $i = \dfrac{7\%}{4} = 1.75\% = 0.0175$; $n = 2.5 \times 4 = 10$
   $S = 2000.00(1.0175)^{10} = 2000.00(1.1894445) = \$2378.89$
   Balance after 6 years:
   $P = 2378.89$; $m = 12$; $i = \dfrac{6.75\%}{12} = 0.5625\% = 0.005625$; $n = 3.5 \times 12 = 42$
   $S = 2378.89(1.005625)^{42} = 2378.89(1.2656528) = \$3010.85$
   After six years the account is worth \$3010.85.

CHAPTER 9

3. Debt value August 1, 2000:

$P = 800.00;\ m = 2;\ i = \dfrac{10\%}{2} = 5\% = 0.05;$

February 1, 1998 – August 1, 2000 contains 2 years 6 months: $n = 2.5 \times 2 = 5$

$S = 800.00(1.05)^5 = 800.00(1.2762816) = \$1021.03$

Debt value November 1, 2003:

$P = 1021.03;\ m = 4;\ i = \dfrac{11\%}{4} = 2.75\% = 0.0275;$

August 1, 2000 – November 1, 2003 contains 3 years 3 months: $n = 3.25 \times 4 = 13$

$S = 1021.03(1.0275)^{13} = 1021.03(1.4228653) = \$1452.79$

The accumulated value of the debt on November 1, 2003 is \$1452.79.

5. $m = 12;\ i = \dfrac{6\%}{12} = 0.5\% = 0.005$

Balance July 1, 1999:

$P = \$1000.00;\ n = 7$

$S = 1000.00(1.005)^7 = 1000.00(1.0355294) = \$1035.53$

Balance November 1, 2000:

$P = 1035.53 + 1000 = 2035.53;\ n = 16$

$S = 2035.53(1.005)^{16} = 2035.63(1.0830712) = \$2204.62$

Balance January 1, 2002:

$P = 2204.62 + 1000.00 = 3204.62;\ n = 14$

$S = 3204.62(1.005)^{14} = 3204.62(1.0723211) = \$3436.38$

The balance in the RRSP account is \$3436.38 on January 1, 2002.

7. $m = 4;\ i = \dfrac{10\%}{4} = 2.5\% = 0.025$

Balance after 9 months:

$P = 4000.00;\ n = 3$

$S = 4000.00(1.025)^3 = 4000.00(1.0768906) = \$4307.56$

Balance after 18 months:

$P = 4307.56 - 1500.00 = 2807.56;\ n = 3$

$S = 2807.56(1.025)^3 = 2807.56(1.0768906) = \$3023.44$

Balance after 27 months:

$P = 3023.44 - 2000.00 = 1023.44;\ n = 3$

$S = 1023.44(1.025)^3 = 1023.44(1.0768906) = \$1102.13$

The size of the final payment is \$1102.13.

9. Balance after 1 year:

$P = 3000.00;\ m = 4;\ i = \dfrac{9\%}{4} = 2.25\% = 0.0225;\ n = 1 \times 4 = 4$

$S = 3000.00(1.0225)^4 = 3000.00(1.0930833) = \$3279.25$

Balance after 2 years:

$P = 3279.25;\ m = 2;\ i = \dfrac{10\%}{2} = 5\% = 0.05;\ n = 1 \times 2 = 2$

$S = 3279.25(1.05)^2 = 3279.25(1.1025) = \$3615.37$

Balance after 4 years:

$P = 3615.37 - 1500.00 = 2115.37;\ i = 0.05;\ n = 2 \times 2 = 4$

$S = 2115.37(1.05)^4 = 2115.37(1.2155063) = \$2571.25$

Balance after 7 years:

$P = 2571.25 - 1500.00 = 1071.25;\ m = 12;\ i = \dfrac{10\%}{12} = 0.8333333\% = 0.0083333;$

$n = 3 \times 12 = 36$

$S = 1071.25(1.0083333)^{36} = 1071.25(1.3481802) = \$1444.24$

The final payment is \$1444.24.

Exercise 9.3

A. 1. $S = 1000.00;\ m = 4;\ i = \dfrac{8\%}{4} = 2.0\% = 0.02;\ n = 7 \times 4 = 28$

$P = \dfrac{S}{(1+i)^n} = S(1+i)^{-n}$

$= 1000.00(1.02)^{-28} = 1000.00(0.5743746) = \$574.37$

3. $S = 600.00;\ m = 12;\ i = \dfrac{8\%}{12} = 0.6666667\% = 0.0066667;\ n = 6 \times 12 = 72$

$P = 600.00(1.0066667)^{-72} = 600.00(0.6197684) = \$371.86$

5. $S = 1200.00;\ m = 12;\ i = \dfrac{9\%}{12} = 0.75\% = 0.0075;\ n = 12 \times 12 = 144$

$P = 1200.00(1.0075)^{-144} = 1200.00(0.3409668) = \$409.16$

7. $S = 900.00;\ m = 4;\ i = \dfrac{6.4\%}{4} = 1.6\% = 0.016;\ n = 9.25 \times 4 = 37$

$P = 900.00(1.016)^{-37} = 900.00(0.5558183) = \$500.24$

B. 1. $S = 1600.00$; $m = 2$; $i = \dfrac{4\%}{2} = 2\% = 0.02$; $n = 4.5 \times 2 = 9$

      $P = 1600.00(1.02)^{-9} = 1600.00(0.8367553) = \$1338.81$

      Compound discount $= 1600.00 - 1338.81 = \$261.19$

3. $S = 1250.00$; $m = 4$; $i = \dfrac{10\%}{4} = 2.5\% = 0.025$; $n = 5 \times 4 = 20$

      $P = 1250.00(1.025)^{-20} = 1250.00(0.6102709) = \$762.84$

5. $S = 5000.00$; $m = 4$; $i = \dfrac{7\%}{4} = 1.75\% = 0.0175$;

      February 1, 2001 – November 1, 2007 contains 6 years 9 months: $n = 6.75 \times 4 = 27$

      $P = 5000.00(1.0175)^{-27} = 5000.00(0.6259948) = \$3129.97$

Exercise 9.4

A. 1.   Maturity value: $FV = PV = 2000.00$; due date 2005-06-30.
      Proceeds:
      Discount period 2002-12-31 to 2005-06-30 contains 2.5 years.

      $FV = 2000.00$; $m = 2$; $i = \dfrac{5\%}{2} = 2.5\% = 0.025$; $n = 2.5(2) = 5$

      $PV = 2000.00(1.025)^{-5} = 2000.00(0.8838543) = \$1767.71$
      Discount $= 2000.00 - 1767.71 = \$232.29$

3.   Maturity value:
      Due date 2007-05-31
      $PV = 1500.00$; $m = 1$; $i = 7\% = 0.07$; $n = 8$

      $FV = 1500.00(1.07)^{8} = 1500.00(1.7181862) = \$2577.28$
      Proceeds:
      Discount period 2004-05-31 to 2007-05-31 contains 3 years.

      $FV = 2577.28$; $m = 2$; $i = \dfrac{8\%}{2} = 4\% = 0.04$; $n = 3 \times 2 = 6$

      $PV = 2577.28(1.04)^{-6} = 2577.28(0.7903145) = \$2036.86$
      Discount $= 2577.28 - 2036.86 = \$540.42$

5. Maturity value:
    Due date 2007-11-01

    $PV = 800.00; \ m = 4; \ i = \dfrac{9\%}{4} = 2.25\% = 0.0225; \ n = 7.75 \times 4 = 31$

    $FV = 800.00(1.0225)^{31} = 800.00(1.9932548) = \$1594.60$

    Proceeds:

    Discount period 2005-11-01 to 2007-11-01 contains 2 years.

    $FV = 1594.60; \ m = 12; \ i = \dfrac{9\%}{12} = 0.75\% = 0.0075; \ n = 2 \times 12 = 24$

    $PV = 1594.60(1.0075)^{-24} = 1594.60(0.8358314) = \$1332.82$

    $Discount = 1594.60 - 1332.82 = \$261.78$

B. 1. $FV = 6000.00; \ m = 4; \ i = \dfrac{6\%}{4} = 1.5\% = 0.015; \ n = \dfrac{54}{12} \times 4 = 18$

    $PV = 6000.00(1.015)^{-18} = 6000.00(0.7649116) = \$4589.47$

    3. Discount period 2002-03-31 to 2005-09-30 contains 3 years 6 months.

    $FV = 1800.00; \ m = 2; \ i = \dfrac{8.5\%}{2} = 4.25\% = 0.0425; \ n = 3.5 \times 2 = 7$

    $PV = 1800.00(1.0425)^{-7} = 1800.00(0.7472528) = \$1345.06$

    5. Maturity value:

    $PV = 3000.00; \ m = 2; \ i = \dfrac{8\%}{2} = 4\% = 0.04; \ n = 5 \times 2 = 10$

    $FV = 3000.00(1.04)^{10} = 3000.00(1.4802443) = \$4440.73$

    Proceeds:

    $FV = 4440.73; \ m = 4; \ i = \dfrac{9\%}{4} = 2.25\% = 0.0225; \ n = \dfrac{21}{12} \times 4 = 7$

    $PV = 4440.73(1.0225)^{-7} = 4440.73(0.8557695) = \$3800.24$

7. Maturity value:
   Due date 2005-06-01

   $PV = 900.00; \; m = 4; \; i = \dfrac{10\%}{4} = 2.5\% = 0.025; \; n = 6 \times 4 = 24$

   $FV = 900.00(1.025)^{24} = 900.00(1.8087260) = \$1627.85$

   Proceeds:

   Discount period 2004-12-01 to 2005-06-01 contains 6 months.

   $FV = 1627.85; \; i = \dfrac{8.5\%}{2} = 4.25\% = 0.0425; \; n = 1$

   $PV = 1627.85(1.0425)^{-1} = 1627.85(0.9592326) = \$1561.49$

9. $FV = 10\,000.00; \; n = 3.5(2) = 7; \; i = \dfrac{5.70\%}{2} = 2.85\%$

   Discounted value $PV = 10\,000.00\left(1.0285^{-7}\right) = 10\,000.00(0.8214288) = \$8214.29$

11. $FV = 5500.00; \; n = 8.5(2) = 17; \; i = \dfrac{5.66\%}{2} = 2.83\%$

    Discounted value $PV = 5500.00\left(1.0283^{-17}\right) = 5500.00(0.6222470) = \$3422.36$

Exercise 9.5

A. 1. $PV = 5000.00; \; i = 0.5\% = 0.005; \; n = 3 \times 12 = 36; \; \text{find FV}$

   $FV = 5000.00(1.005)^{36} = 5000.00(1.1966805) = \$5983.40$

3. $FV = 3400.00; \; i = 5\% = 0.05; \; n = 3 \times 2 = 6; \; \text{find PV}$

   $PV = 3400.00(1.05)^{-6} = 3400.00(0.7462154) = \$2537.13$

5. $PV_1 = 800.00; \; i = 0.7916667\% = 0.0079167; \; n = 18$

   $FV_1 = 800.00(1.0079167)^{18} = 800.00(1.1525069) = \$922.01$

   $PV_2 = 700.00; \; i = 0.0079167; n = 9$

   $FV_2 = 700.00(1.0079167)^{9} = 700.00(1.0735487) = \$751.48$

   $FV = FV_1 + FV_2 = 922.01 + 751.48 = \$1673.49$

7. $FV_1 = 400.00;\ i = 5.5\% = 0.055;\ n = 3 \times 2 = 6$

$PV_1 = 400.00(1.055)^{-6} = 400.00(0.7252458) = \$290.10$

$FV_2 = 600.00;\ i = 0.055;\ n = 5 \times 2 = 10$

$PV_2 = 600.00(1.055)^{-10} = 600.00(0.5854306) = \$351.26$

$PV = PV_1 + PV_2 = 290.10 + 351.26 = \$641.36$

9. $PV_1 = 800.00;\ i = 1.875\% = 0.01875;\ n = 4$

$E_1 = FV_1 = 800.00(1.01875)^4 = 800.00(1.0771359) = \$861.71$

$FV_2 = 1400.00(1.065)^3 = 1400.00(1.2079496) = 1691.13;$

$\qquad i = 0.01875;\ n = 2 \times 4 = 8$

$E_2 = PV_2 = 1691.13(1.01875)^{-8} = 1691.13(0.8619043) = \$1457.59$

$E = E_1 + E_2 = 861.71 + 1457.59 = \$2319.30$

B. 1. $E_1 = 2000.00(1.105)^7 = 2000.00(2.0115737) = \$4023.15$

$E_2 = 2000.00(1.105)^3 = 2000.00(1.3492326) = \$2698.47$

$E_3 = 2000.00(1.105)^5 = 2000.00(1.6474468) = \$3294.89$

$E_1 + E_2 = E_3 + x$

$4023.15 + 2698.47 = 3294.89 + x$

$x = 3426.73$

The second payment is \$3426.73.

3. $E_1 = 800.00(1.06)^{12} = 800.00(2.0121965) = \$1609.76$

$E_2 = 1000.00(1.06)^{-2} = 1000.00(0.8899964) = \$890.00$

$E_3 = \$x$

$E_4 = x(1.06)^{-8} = x(0.6274124)$

$E_1 + E_2 = E_3 + E_4$

$1609.76 + 890.00 = x + 0.6274124x$

$2499.76 = 1.6274124x$

$x = 1536.03$

The size of the two equal payments is \$1536.03.

5. $E_1 = 900.00(1.0075)^{-3} = 900.00(0.9778333) = \$880.05$

$\quad\quad 800.00(1.0275)^{10} = 800.00(1.3116510) = 1049.32$

$E_2 = 1049.32(1.0075)^{-30} = 1049.32(0.7991869) = \$838.60$

$E_3 = \$x$

$E_4 = x(1.0075)^{-36} = 0.7641490x$

$\quad\quad E_1 + E_2 = E_3 + E_4$

$880.05 + 838.60 = x + 0.7641490x$

$\quad\quad 1718.65 = 1.7641490x$

$\quad\quad\quad\quad x = 974.21$

The size of the two equal payments is $974.21.

C. 1. (a) $E = 4000.00(1.07)^{-5} = 4000.00(0.7129862) = \$2851.94$

(b) $E = 4000.00(1.07)^{-3} = 4000.00(0.8162979) = \$3265.19$

(c) $E = \$4000.00$

(d) $E = 4000.00(1.07)^{5} = 4000.00(1.4025517) = \$5610.21$

3. Focal date is 4 years from now.

$E_1 = 2000.00(1.05)^{8} \doteq 2000.00(1.4774554) = \$2954.91$

$E_2 = 2000.00(1.05)^{2} = 2000.00(1.1025) = \$2205.00$

$E_3 = 2000.00(1.05)^{-4} = 2000.00(0.8227025) = \$1645.41$

$x = E_1 + E_2 + E_3$

$\quad = 2954.91 + 2205.00 + 1645.41$

$\quad = \$6805.32$

The single payment four years from now is $6805.32.

5. Focal date is 15 months from now.

$E_1 = 400.00(1.005)^{15} = 400.00(1.0776827) = \$431.07$

$E_2 = \left[700.00(1.00375)^{8} \right](1.005)^{7}$

$\quad = 700.00(1.0303967)(1.005)^{7}$

$\quad = 721.28(1.0355294) = \$746.91$

$E_3 = 500.00(1.005)^{9} = 500.00(1.0459106) = \$522.96$

$\quad\quad E_1 + E_2 = E_3 + x$

$431.07 + 746.91 = 655.02 + x$

$\quad\quad\quad\quad x = 655.02$

The final payment is $655.02.

7. Let the size of the equal payments be $x and the focal date be now.

$$8000.00(1.04)^2 = x + x(1.04)^{-2} + x(1.04)^{-4} + x(1.04)^{-6}$$
$$8000.00(1.0816) = x + 0.9245562x + 0.8548042x + 0.7903145x$$
$$8652.80 = 3.5696749x$$
$$x = 2423.97$$

The size of the equal payments is $2423.97.

9. (a) Focal date is 5 years from now.

$$500.00(1.11)^4 + 500.00(1.11)^3 + 500.00(1.11)^2 + 500.00(1.11) + 500.00 = x$$
$$x = 500.00(1.5180704 + 1.367631 + 1.2321 + 1.11 + 1)$$
$$x = 500.00(6.2278014)$$
$$x = 3113.90$$

The single payment is $3113.90.

(b) Focal date is now.

$$x = 500.00(1.11)^{-1} + 500.00(1.11)^{-2} +$$
$$500.00(1.11)^{-3} + 500.00(1.11)^{-4} + 500.00(1.11)^{-5}$$
$$= 500.00(0.9009009 + 0.8116224 + 0.7311914 +$$
$$0.6587310 + 0.5934513)$$
$$= 500.00(3.6958970)$$
$$= 1847.95$$

The single payment now is $1847.95.

11. (a) Let the size of the equal payments be $x and the focal date be 4 years from now.

$$3000.00 = x(1.10)^3 + x(1.10)^2 + x(1.10) + x$$
$$3000.00 = x(1.10^3 + 1.10^2 + 1.10 + 1)$$
$$3000.00 = x(1 + 1.10 + 1.21 + 1.331)$$
$$3000.00 = x(4.641)$$
$$x = \$646.41$$

(b) Focal date is now.

$$3000.00 = x(1.10)^{-1} + x(1.10)^{-2} + x(1.10)^{-3} + x(1.10)^{-4}$$
$$3000.00 = x(1.10^{-1} + 1.10^{-2} + 1.10^{-3} + 1.10^{-4})$$
$$3000.00 = x(0.9090909 + 0.8264463 + 0.7513148 + 0.6830135)$$
$$3000.00 = 3.1698655x$$
$$x = \$946.41$$

Review Exercise

1. (a) $FV = 500.00(1.06)^{15} = 500.00(2.3965582) = \$1198.28$

    (b) $FV = 500.00(1.015)^{60} = 500.00(2.4432198) = \$1221.61$

    (c) $FV = 500.00(1.005)^{180} = 500.00(2.4540936) = \$1227.05$

3. (a) $FV = 2000.00(1.0375)^{10} = 2000.00(1.4450439) = \$2890.09$

    (b) Interest $= 2890.09 - 2000.00 = \$890.09$

5. (a) $PV = 1800.00(1.02)^{62} = 1800.00(3.4135844) = \$6144.45$
        Interest $= 6144.45 - 1800.00 = \$4344.45$

    (b) $FV = 1250.00(1.0054167)^{180} = 1250.00(2.6442166) = \$3305.27$
        Interest $= 3305.27 - 1250.00 = \$2055.27$

7. Accumulated value after 2 years
    $= 20\,000.00(1.05)^{4} = 20\,000.00(1.2155063) = \$24\,310.13$
    Balance after payment $= 24\,310.13 - 8000.00 = \$16\,310.13$
    Accumulated value after 3.5 years
    $= 16\,310.13(1.05)^{3} = 16\,310.13(1.157625) = \$18\,881.01$
    Balance after payment $= 18\,881.01 - 10\,000.00 = \$8881.01$
    Accumulated value one year later
    $= 8881.01(1.05)^{2} = 8881.01(1.1025) = \$9791.31$

9. Discount period 2002-09-30 to 2008-06-30 contains 5 years 9 months
    $PV = 1500.00(1.025)^{-23} = 1500.00(0.5666972) = \$850.05$

11. Balance 1 year after 1st deposit

$= 2000.00(1.015)^4 = 2000.00(1.0613636) = \$2122.73$

Balance 2 years after 1st deposit

$= (2122.73 + 2000.00)(1.015)^4 = 4122.73(1.0613636) = \$4375.72$

Balance 4 years after 1st deposit

$= (4375.72 + 2000.00)(1.015)^8 = 6375.72(1.1264926) = \$7182.20$

13. Balance after 15 months

$= 8000.00(1.02)^5 - 3000.00$

$= 8000.00(1.1040808) - 3000.00$

$= 8832.65 - 3000.00$

$= \$5832.65$

Balance after 24 months

$= 5832.65(1.02)^3 = 5832.65(1.0612080) = \$6189.65$

Balance after 30 months

$= 6189.65(1.0225)^2 - 4000.00$

$= 6189.65(1.045506) - 4000.00$

$= \$6471.32 - 4000.00$

$= \$2471.32$

Balance after 4 years

$= 2471.32(1.0225)^6 = 2471.32(1.1428254) = \$2824.29$

15. $PV = 6000.00(1.0041667)^{-180} = 6000.00(0.4731003) = \$2838.60$

17. Due date May 1, 2007

Maturity value $= 1750.00(1.02)^{20} = 1750.00(1.4859474) = \$2600.41$

Discount period 2003-08-01 to 2007-05-01 contains 3 years 9 months

Proceeds $= 2600.41(1.015)^{-15} = 2600.41(0.7998515) = \$2079.94$

19. Due date 2013-06-01

Maturity value $= 40\,000(1.06)^{30} = 40\,000.00(5.7434912) = \$229\,739.65$

Discount period 2002-09-01 to 2013-06-01 contains 10 years 9 months

Proceeds $= 229\,739.65(1.0275)^{-43} = 229\,739.65(0.3114449) = \$71\,551.24$

21. $FV = 100\,000.00;\ n = 15.5(2) = 31;\ i = \dfrac{5.15\%}{2} = 2.575\% = 0.02575$

Discounted value $PV = 100\,000.00\left(1.02575^{-31}\right) = 100\,000.00(0.4546872) = \$45\,468.72$

23. $FV = 2750.00;\ n = 7(2) = 14;\ i = \dfrac{6.14\%}{2} = 3.07\% = 0.0307$

Discounted value $PV = 2750.00\left(1.0307^{-14}\right) = 2750.00(0.6548595) = \$1800.86$

25. Let the single payment be $\$x$ and the focal date be two years from now.
$$x = 1000.00(1.024)^6 + 1200.00(1.024)^2 + 1500.00(1.024)^{-2}$$
$$= 1000.00(1.1529215) + 1200.00(1.0485760) + 1500.00(0.9536743)$$
$$= 1152.92 + 1258.29 + 1430.51$$
$$= \$3841.72$$

27. Let the size of the equal payments be $\$x$ and the focal date be now.
Accumulated value of $3000 due in 2 years
$$= 3000.00(1.055)^4 = 3000.00(1.2388247) = \$3716.47$$
Accumulated value of $2500 due in 15 months
$$= 2500.00(1.0225)^5 = 2500.00(1.1176777) = \$2794.19$$
$$3716.47(1.007)^{-24} + 2794.19(1.007)^{-15} = x + x(1.007)^{-18}$$
$$3716.47(0.8458487) + 2794.19(0.9006539) = x + 0.8820019x$$
$$3143.57 + 2516.60 = 1.8820019x$$
$$x = \$3007.53$$

29. Let the size of the equal payments be $\$x$ and the focal date be one year from now.
$$2600.00(1.048)^4 + 2400.00(1.048)^{-2} = x + x(1.048)^{-6}$$
$$2600.00(1.2062717) + 2400.00(0.9104947) = x + 0.7548007x$$
$$3136.31 + 2185.19 = 1.7548007x$$
$$5321.50 = 1.7548007x$$
$$x = \$3032.54$$

1. $FV = 3300.00$; $i = 1\%$; $n = 44$

   $PV = S(1 + i)^{-n} = 3300.00(1.01)^{-44} = 3300.00(0.6454455) = \$2129.97$

3. $i = 3.5\%$; $n = 29$

   $(1 + i)^n = 1.035^{29} = 2.7118780$

5. Let the size of the single payment be $\$x$ and the focal date be 1 year from now.

   $x = 4000.00(1.035)^2 + 4000.00(1.035)^{-8} + 3000.00(1.035)^{-10}$

   $ = 4000.00(1.071225) + 4000.00(0.7594116) + 3000.00(0.7089188)$

   $ = 4284.90 + 3037.65 + 2126.76$

   $ = \$9449.31$

7. $FV = 5900.00$; $i = 3.75\%$; $n = 30$

   $PV = 5900.00(1.0375)^{-30} = 5900.00(0.3314033) = \$1955.28$

9. Maturity value of $600 due in 9 months:

   $PV = 600.00$; $i = 0.875\%$; $n = 9$

   $FV = 600.00(1.00875)^9 = 600.00(1.0815633) = \$648.94$

   Let the size of the final payment be $\$x$ and the focal date be 24 months from now.

   $800.00(1.02375)^8 + 648.94(1.02375)^5 = 800.00(1.02375)^6 + x$

   $800.00(1.2065667) + 648.94(1.1245262) = 800.00(1.1512337) + x$

   $965.25 + 729.75 = 920.99 + x$

   $x = \$774.01$

11. Balance after 2 years: $PV = 5000.00$; $i = 5\%$; $n = 4$

    $FV = 5000.00(1.05)^4 = 5000.00(1.2155063) = \$6077.53$

    Balance $= 6077.53 - 2000.00 = \$4077.53$

    Balance after 3 years: $PV = 4077.53$; $i = 5\%$; $n = 2$

    $FV = 4077.53(1.05)^2 = 4077.53(1.1025) = \$4495.48$

    Balance $= 4495.48 - 2500.00 = \$1995.48$

    Balance after 5 years: $PV = 1995.48$; $i = 5\%$; $n = 4$

    $FV = 1995.48(1.05)^4 = 1995.48(1.2155063) = \$2425.52$

10 Compound interest—Further topics

Exercise **10.1**

A. 1. (a) PV = 2500.00; i = 7% = 0.07; n = 7.5

   FV = PV$(1 + i)^n$ = 2500.00$(1.07)^{7.5}$ = 2500.00(1.6610333) = \$4152.58

(b) PV = 400.00; i = 2.25% = 0.0225;

   $n = 3 \times 4 + \dfrac{8}{12} \times 4 = 12 + 2\dfrac{2}{3} = 14.666667$

   FV = 400.00$(1.0225)^{14.666667}$ = 400.00(1.3858896) = \$554.36

(c) PV = 1300.00; i = 2.5% = 0.025; $n = 9 \times 2 + \dfrac{3}{12} \times 2 = 18 + \dfrac{1}{2} = 18.5$

   FV = 1300.00$(1.025)^{18.5}$ = 1300.00(1.5790341) = \$2052.74

(d) PV = 4500.00; i = 0.2916667% = 0.002916667; n = 7.5

   FV = 4500.00$(1.002916667)^{7.5}$ = 4500.00(1.0220835) = \$4599.38

3. (a) FV = 1500.00; i = 4.5% = 0.045; $n = 15\dfrac{9}{12} = 15.75$

   PV = FV$(1 + i)^{-n}$ = 1500.00$(1.045)^{-15.75}$ = 1500.00(0.4999406) = \$749.91

(b) FV = 900.00; i = 2.75% = 0.0275; $n = 8 \times 2 + \dfrac{10}{12} \times 2 = 16 + \dfrac{10}{6} = 17\dfrac{2}{3}$

   PV = 900.00$(1.0275)^{-17.666667}$ = 900.00(0.6192333) = \$557.31

(c) FV = 6400.00; i = 1.75% = 0.0175; $n = 5 \times 4 + \dfrac{7}{12} \times 4 = 20 + \dfrac{7}{3} = 22\dfrac{1}{3}$

   PV = 6400.00$(1.0175)^{-22.333333}$ = 6400.00(0.6787836) = \$4344.21

(d) FV = 7200.00; i = 0.5% = 0.005; n = 21.5

   PV = 7200.00$(1.005)^{-21.5}$ = 7200.00(0.89831711) = \$6467.88

5. (a) Due date: 2008-04-01
        Discount period: 2003-08-01 to 2008-04-01: 4 years 8 months

        $FV = 5000.00; \; i = 0.05; \; n = 4\dfrac{8}{12} = 4.6666667$

        $PV = 5000.00(1.05)^{-4.6666667} = 5000.00(0.7963731) = \$3981.87$

    (b) Due date: 2007-08-31
        Discount period: 2004-06-30 to 2007-08-31: 3 years 2 months

        $FV = 900.00; \; i = 0.01; \; n = 4\left(3\dfrac{2}{12}\right) = 12.6666667$

        $PV = 900.00(1.01)^{-12.6666667} = 900.00(0.8815818) = \$793.42$

    (c) Due date: 2006-03-31
        Discount period: 2003-10-31 to 2006-03-31: 2 years 5 months

        $FV = 3200.00; \; i = 0.02; \; n = 4\left(2\dfrac{5}{12}\right) = 9.6666667$

        $PV = 3200.00(1.02)^{-9.6666667} = 3200.00(0.8257812) = \$2642.50$

    (d) Due date: 2006-10-01
        Discount period: 2002-12-01 to 2006-10-01: 3 years 10 months

        $FV = 1450.00; \; i = 0.03; \; n = 2\left(3\dfrac{10}{12}\right) = 7.6666667$

        $PV = 1450.00(1.03)^{-7.6666667} = 1450.00(0.7972257) = \$1155.98$

    (e) Due date: 2007-09-30
        $PV = 780.00; \; i = 0.08; \; n = 10$
        Maturity value $= 780.00(1.08)^{10} = 780.00(2.1589250) = \$1683.96$
        Discount period: 2001-04-30 to 2007-09-30: 6 years 5 months

        $FV = 1683.96; \; i = 0.02; \; n = 4\left(6\dfrac{5}{12}\right) = 25.6666667$

        $\text{Proceeds} = 1683.96(1.02)^{-25.6666667}$
        $\qquad\qquad = 1683.96(0.6015369) = \$1012.96$

    (f) Due date: 2008-02-01
        $PV = 2100.00; \; i = 0.005; \; n = 144$
        Maturity value $= 2100.00(1.005)^{144} = 2100.00(2.0507508) = \$4306.58$
        Discount period: 2003-07-01 to 2008-02-01: 4 years 7 months

        $FV = 4306.58; \; i = 0.035; \; n = 2\left(4\dfrac{7}{12}\right) = 9.1666667$

        $\text{Proceeds} = 4306.58(1.035)^{-9.1666667}$
        $\qquad\qquad = 4306.58(0.7295361) = \$3141.81$

CHAPTER 10

(g) Due date: 2007-11-01
PV = 1850.00; i = 0.025; n = 20
Maturity value = $1850.00(1.025)^{20}$ = 1850.00(1.6386164) = \$3031.44
Discount period: 2004-10-01 to 2007-11-01: 3 years 1 month

FV = 3031.44; i = 0.045; $n = 2\left(3\frac{1}{12}\right)$ = 6.1666667

Proceeds = $3031.44(1.045)^{-6.1666667}$
= 3031.44(0.7622830) = \$2310.82

(h) Due date: 2006-01-31
PV = 3400.00; i = 0.0075; n = 84
Maturity value = $3400.00(1.0075)^{84}$ = 3400.00(1.8732020) = \$6368.89
Discount period: 2003-12-31 to 2006-01-31: 2 years 1 month

FV = 6368.89; i = 0.01875; $n = 4\left(2\frac{1}{12}\right)$ = 8.3333333

Proceeds = $6368.89(1.01875)^{-8.3333333}$
= 6368.89(0.8565837) = \$5455.49

B. 1. PV = 5000.00; i = 4.875% = 0.04875; $n = 2\left(5\frac{10}{12}\right)$ = 11.6666667

FV = $5000.00(1.04875)^{11.6666667}$
= 5000.00(1.7425010)
= 8712.50
I = 8712.50 − 5000.00 = \$3712.50

3. Interest period 2000-08-01 to 2005-06-01 contains 4 years 10 months.
PV = 600.00; i = 2.5% = 0.025; $n = 4 \times 2 + \frac{10}{12} \times 2 = 8 + \frac{10}{6} = 9\frac{2}{3}$
Maturity value = $600.00(1.025)^{9.6666667}$
= 600.00(1.2695916)
= \$761.75

5. PV = 8000.00; i = 10.8% = 0.108; $n = 7\frac{5}{12}$ = 7.4166667
FV = $8000.00(1.108)^{7.4166667}$
= 8000.00(2.1396196)
= \$17116.96

7. $FV = 3000.00$; $i = 4.5\% = 0.045$; $n = 2\left(8\dfrac{8}{12}\right) = 17.3333333$

 $\begin{aligned} PV &= 3000.00(1.045)^{-17.3333333} \\ &= 3000.00(0.4662845) \\ &= \$1398.85 \end{aligned}$

9. $FV = 10\,000.00$; $i = 0.75\% = 0.0075$; $n = 22.5$

 $\text{Proceeds} = 10\,000.00(1.0075)^{-22.5} = 10\,000.00(0.8452521) = \$8452.52$

11. $FV = 3800.00$; $i = 7.5\% = 0.075$; $n = 6\dfrac{8}{12} = 6.6666667$

 $\begin{aligned} PV &= 3800.00(1.075)^{-6.6666667} \\ &= 3800.00(0.6174620) \\ &= \$2346.36 \end{aligned}$

13. Discount period $= 48 - 32 = 16$ months

 $FV = 3750.00$; $i = 2.75\% = 0.0275$; $n = \dfrac{16}{12} \times 2 = \dfrac{16}{6} = 2\dfrac{2}{3}$

 $PV = 3750.00(1.0275)^{-2.6666667} = 3750.00(0.9302117) = \$3488.29$

15. Due date: 2006-08-01
 Discount period: 2002-04-01 to 2006-08-01: 4 years 4 months

 $FV = 4500.00$; $i = 0.065$; $n = 4\dfrac{4}{12} = 4.3333333$

 $\begin{aligned} \text{Proceeds} &= 4500.00(1.065)^{-4.3333333} \\ &= 4500.00(0.7611759) \\ &= 3425.29 \end{aligned}$

 $\text{Discount} = 4500.00 - 3425.29 = \$1074.71$

17. $FV = 75\,000.00$; $i = 3.68\%$
 The discount period November 20, 2001 to July 20, 2010 contains 8 years, 8 months:

 $n = 2\left(8\dfrac{8}{12}\right) = 17.3333333$

 $\begin{aligned} PV &= 75\,000.00\left(1.0368^{-17.3333333}\right) = 75\,000.00(0.5345072) \\ &= \$40\,088.04 \end{aligned}$

19. Due date: 2007-12-01
    PV = 1750.00; i = 6.5% = 0.065; n = 6
    Maturity value = $1750.00(1.065)^6$ = 1750.00(1.4591423) = \$2553.50
    Discount period: 2004-03-01 to 2007-12-01: 3 years 9 months

    FV = 2553.50; i = 3.5% = 0.035; $n = 3 \times 2 + \dfrac{9}{12} \times 2 = 6 + \dfrac{3}{2} = 7.5$

    PV = $2553.50(1.035)^{-7.5}$ = 2553.50(0.7725870) = \$1972.80

21. PV = 2650.00; i = 2.25% = 0.0225; n = 28
    Maturity value = $2650.00(1.0225)^{28}$ = 2650.00(1.8645450) = \$4941.04

    FV = 4941.04; i = 0.04; $n = 2\left(4\dfrac{7}{12}\right) = 9.1666667$

    Proceeds = $4941.04(1.04)^{-9.1666667}$
     = 4941.04(0.6980091)
     = 3448.89

    Discount = 4941.04 − 3448.89 = \$1492.15

Exercise 10.2

A. 1. (a) PV = 2600.00; FV = 6437.50; i = 7% = 0.07
    $$6437.50 = 2600.00(1.07)^n$$
    $$1.07^n = 2.4759615$$
    $$n \ln 1.07 = \ln 2.4759615$$
    $$0.0676586n = 0.9066288$$
    $$n = 13.4 \,(\text{years})$$

    (b) PV = 1240.00; FV = 1638.40; i = 1% = 0.01
    $$1638.40 = 1240.00(1.01)^n$$
    $$1.01^n = 1.3212903$$
    $$n \ln 1.01 = \ln 1.3212903$$
    $$0.00995033n = 0.2786088$$
    $$n = 27.999956$$
    $$n = 28 \,(\text{quarters})$$

(c)   PV = 560.00; FV = 1350.00; $i = 0.75\% = 0.0075$

$$1350.00 = 560.00(1.0075)^n$$

$$1.0075^n = 2.4107143$$

$$n\ln 1.0075 = \ln 2.4107143$$

$$0.00747201n = 0.879923$$

$$n = 117.763$$

$$n = 117.8 \text{ (months)}$$

(d)   PV = 3480.00; FV = 4762.60; $i = 4\% = 0.04$

$$4762.60 = 3480.00(1.04)^n$$

$$1.04^n = 1.3685632$$

$$n\ln 1.04 = \ln 1.3685632$$

$$0.0392207n = 0.3137614$$

$$n = 7.9998929$$

$$n = 8 \text{ (half - years)}$$

(e)   PV = 950.00; FV = 1900.00; $i = 1.875\% = 0.01875$

$$1900.00 = 950.00(1.01875)^n$$

$$1.01875^n = 2$$

$$n\ln 1.01875 = \ln 2$$

$$0.0185764n = 0.6931472$$

$$n = 37.3133$$

$$n = 37.313 \text{ (quarters)}$$

(f)   P V = 1300.00; FV = 3900.00; $i = 3.0 = 0.03$

$$3900.00 = 1300.00(1.03)^n$$

$$1.03^n = 3$$

$$n\ln 1.03 = \ln 3$$

$$0.0295588n = 1.0986123$$

$$n = 37.1680188$$

$$n = 37.168 \text{ (half - years)}$$

B.   1.   PV = 400.00; FV = 760.00; $i = 3.5\% = 0.035$

$$760.00 = 400.00(1.035)^n$$

$$1.035^n = 1.9$$

$$n\ln 1.035 = \ln 1.9$$

$$0.0344014n = 0.6418539$$

$$n = 18.6577843$$

$$n = 18.658 \text{ (half - years)}$$

It takes about 9.329 years.

3.   Let PV = 1; FV = 4; i = 2.0% = 0.02; m = 4

$$1.02^n = 4$$
$$n \ln 1.02 = \ln 4$$
$$0.0198026n = 1.3862944$$
$$n = 70.005676 \text{ (quarters)}$$

Number of years = $\dfrac{1}{4} \times 70.005676 = 17.501$ years.

5.   PV = 800.00; I = 320.00; FV = 1120.00; i = 0.5% = 0.005; m = 12

$$800.00(1.005)^n = 1120.00$$
$$1.005^n = 1.4$$
$$n \ln 1.005 = \ln 1.4$$
$$0.0049875n = 0.3364722$$
$$n = 67.463098 \text{ months}$$

Number of years = $\dfrac{67.463098}{12} = 5.622$ years.

7.   PV = 1000.00; I = 157.63; FV = 1157.63; i = 5% = 0.05

$$1157.63 = 1000.00(1.05)^n$$
$$1.05^n = 1.15763$$
$$n \ln 1.05 = \ln 1.15763$$
$$0.0487902n = 0.1463748$$
$$n = 3.0000861 \text{ (half - years)}$$

1.5 years after May 1, 2000 is November 1, 2001.

9.   PV = 1375.07; FV = 1500.00; i = 1.25% = 0.0125; m = 4

$$1500.00 = 1375.07(1.0125)^n$$
$$1.0125^n = 1.0908536$$
$$n \ln 1.0125 = \ln 1.0908536$$
$$0.0124225n = 0.0869605$$
$$n = 7.0002415 \text{ (quarters)}$$

The discount date is $7 \times 3 = 21$ months before the due date.

11.   Let the focal date be now; $i = 0.75\% = 0.0075$; $m = 12$

$$4000.00 + 5000.00(1.0075)^{-36} + 6000.00(1.0075)^{-60} = 15\,000.00(1.0075)^{-n}$$

$$4000.00 + 5000.00(0.7641490) + 6000.00(0.6386997) = 15\,000.00(1.0075)^{-n}$$

$$4000.00 + 3820.74 + 3832.20 = 15\,000.00(1.0075)^{-n}$$

$$11\,652.94 = 15\,000(1.0075)^{-n}$$

$$1.0075^{-n} = 0.7768627$$

$$-n\ln 1.0075 = \ln 0.7768627$$

$$-0.0074720n = -0.2524916$$

$$n = 33.791702 \text{ (months)}$$

$$n = 2.8159752 \text{ (years)}$$

The equated date is 2 years and 298 days from now.

13.   Let the focal date be now; $i = 4.875\% = 0.04875$; $m = 2$

$$6000.00(1.04875)^{-2} + 6000.00(1.04875)^{-6} + 6000.00(1.04875)^{-10} = 8000.00(1.04875)^{-4}$$
$$+ 10\,000(1.04875)^{-n}$$

$$6000.00(0.9091929) + 6000.00(0.7515678) + 6000.00(0.6212698) = 8000.00(0.8266318)$$
$$+ 10\,000(1.04875)^{-n}$$

$$5455.16 + 4509.41 + 3727.62 = 6613.05$$
$$+ 10\,000.00(1.04875)^{-n}$$

$$7079.14 = 10\,000(1.04875)^{-n}$$

$$1.04875^{-n} = 0.707914$$

$$-n\ln 1.04875 = \ln 0.707914$$

$$-0.0475990n = -0.3454327$$

$$n = 7.257142 \text{ (half - years)}$$

$$n = 3.628571 \text{ (years)}$$

The second payment should be made in 3 years and 230 days.

Exercise 10.3

1.   (a) $PV = 1400.00$; $FV = 1905.21$; $n = 7$; $m = 1$

$$1905.21 = 1400.00(1 + i)^7$$

$$(1 + i)^7 = 1.3608643$$

$$(1 + i) = 1.3608643^{\frac{1}{7}}$$

$$1 + i = 1.0450003$$

$$i = 0.0450003$$

$$i = 4.500\%$$

The nominal annual rate is 4.5%.

(b) PV = 2350.00; FV = 3850.00; $n = 20$; $m = 4$

$$3850.00 = 2350.00(1 + i)^{20}$$

$$(1 + i)^{20} = 1.6382979$$

$$(1 + i) = 1.6382979^{\frac{1}{20}}$$

$$1 + i = 1.02499$$

$$i = 0.025$$

$$i = 2.5\%$$

The nominal annual rate is $2.5\% \times 4 = 10.0\%$.

(c) PV = 690.00; FV = 1225.00; $n = 72$; $m = 12$

$$1225.00 = 690.00(1 + i)^{72}$$

$$(1 + i)^{72} = 1.7753623$$

$$1 + i = 1.7753623^{\frac{1}{72}}$$

$$1 + i = 1.0080041$$

$$i = 0.8\%$$

The nominal annual rate is $0.8\% \times 12 = 9.6\%$.

(d) PV = 1240.00; FV = 2595.12; $n = 24$; $m = 2$

$$2595.12 = 1240.00(1 + i)^{24}$$

$$(1 + i)^{24} = 2.0928387$$

$$(1 + i) = 2.0928387^{\frac{1}{24}}$$

$$1 + i = 1.03125007$$

$$i = 3.125007\%$$

The nominal annual rate is $3.125\% \times 2 = 6.25\%$.

(e) PV = 3160.00; FV = 5000.00; $n = 19$; $m = 4$

$$5000.00 = 3160.00(1 + i)^{19}$$

$$(1 + i)^{19} = 1.5822785$$

$$(1 + i) = 1.5822785^{\frac{1}{19}}$$

$$1 + i = 1.0244448$$

$$i = 2.44448\%$$

The nominal annual rate is $2.44448\% \times 4 = 9.778\%$.

(f)   PV = 900.00; FV = 1200.00; $n = 44$; $m = 12$

$$1200.00 = 900.00(1 + i)^{44}$$

$$(1 + i)^{44} = 1.3333333$$

$$(1 + i) = 1.3333333^{\frac{1}{44}}$$

$$1 + i = 1.00655965$$

$$i = 0.655965\%$$

The nominal annual rate is $0.655965\% \times 12 = 7.872\%$.

3. (a) $i = 4.75\% = 0.0475$; $m = 2$

$$f = (1 + i)^{m} - 1 = (1.0475)^{2} - 1 = 1.0972563 - 1 = 0.0972563 = 9.726\%$$

(b) $i = 2.625\% = 0.02625$; $m = 4$

$$f = (1.02625)^{4} - 1 = 1.1092072 - 1 = 0.1092072 = 10.921\%$$

(c) $i = 0.4166667\% = 0.0041667$; $m = 12$

$$f = (1.0041667)^{12} - 1 = 1.0511623 - 1 = 0.0511623 = 5.116\%$$

(d) $i = 0.6\% = 0.006$; $m = 12$

$$f = (1.006)^{12} - 1 = 1.0744242 - 1 = 0.0744242 = 7.442\%$$

(e) $i = 0.9\% = 0.009$; $m = 4$

$$f = (1.009)^{4} - 1 = 1.0364889 - 1 = 0.0364889 = 3.649\%$$

(f) $i = 4.1\% = 0.041$; $m = 2$

$$f = (1.041)^{2} - 1 = 1.0836810 - 1 = 0.0836810 = 8.368\%$$

B. 1.   PV = 420.00; FV = 1000.00; $n = 38$; $m = 4$

$$1000.00 = 420.00(1 + i)^{38}$$

$$(1 + i)^{38} = 2.3809524$$

$$(1 + i) = 2.3809524^{\frac{1}{38}}$$

$$1 + i = 1.0230915$$

$$i = 2.30915\%$$

The nominal annual rate is $2.30915\% \times 4 = 9.237\%$.

CHAPTER 10

3.   Let PV = 1; FV = 2

(a) $n = 27$; $m = 4$
$$(1 + i)^{27} = 2$$
$$1 + i = 2^{\frac{1}{27}}$$
$$1 + i = 1.0260045$$
$$i = 2.60045\%$$
The nominal annual rate is $2.60045 \times 4 = 10.402\%$.

(b) $n = 110$; $m = 12$
$$(1 + i)^{110} = 2$$
$$(1 + i) = 2^{\frac{1}{110}}$$
$$(1 + i) = 1.0063212$$
$$i = 0.63212\%$$
The nominal annual rate is $0.63212\% \times 12 = 7.585\%$.

5.   PV = 100.00; FV = 150.00; $n = 24$; $m = 4$
$$150.00 = 100.00(1 + i)^{24}$$
$$(1 + i)^{24} = 1.5$$
$$(1 + i) = 1.5^{\frac{1}{24}}$$
$$1 + i = 1.0170379$$
$$f = (1 + i)^{m} - 1$$
$$= 1.0170379^{4} - 1$$
$$= 1.0699132 - 1$$
$$= 0.0699132$$
$$= 6.991\%$$

7. $f = 9.25\% = 0.0925$; $m = 4$
$$f = (1 + i)^{m} - 1$$
$$0.0925 = (1 + i)^{4} - 1$$
$$(1 + i)^{4} = 1.0925$$
$$(1 + i) = 1.0925^{\frac{1}{4}}$$
$$1 + i = 1.0223636$$
$$i = 2.23636\%$$
The nominal annual rate is $2.23636\% \times 4 = 8.945\%$.

9. Value of \$1 in one year at 7.5% compounded semi-annually.

$$FV_1 = 1.0375^2 = 1.0764063$$

Value of \$1 in one year at a monthly rate i, $FV_2 = (1 + i)^{12}$

for $FV_1 = FV_2$

$$1.0764063 = (1 + i)^{12}$$
$$1 + i = 1.0764063^{\frac{1}{12}}$$
$$1 + i = 1.0061545$$
$$i = 0.61545\%$$

The nominal rate is $0.61545\% \times 12 = 7.385\%$.

Exercise 10.4

A. 1. (a) $PV = 400.00$; $n = 6$; $j = 7\% = 0.07$

$$FV = PVe^{nj}$$
$$= 400.00e^{(6)(0.07)}$$
$$= 400.00e^{0.42}$$
$$= 400.00(1.5219616)$$
$$= \$608.78$$

(b) $PV = 2700.00$; $n = 12$; $j = 11.5\% = 0.115$

$$FV = 2700.00e^{(12)(0.115)}$$
$$= 2700.00e^{1.38}$$
$$= 2700.00(3.9749016)$$
$$= \$10\,732.23$$

(c) $PV = 1800.00$; $j = 5.5\% = 0.055$; $n = 4\frac{7}{12}$

$$FV = 1800.00e^{(0.055)(4.5833333)}$$
$$= 1800.00e^{0.2520833}$$
$$= 1800.00(1.2867033)$$
$$= \$2316.07$$

(d) $PV = 3700.00$; $j = 10.4\% = 0.104$; $n = 3\frac{10}{12}$

$$FV = 3700.00e^{(0.104)(3.8333333)}$$
$$= 3700.00e^{0.3986667}$$
$$= 3700.00(1.4898369)$$
$$= \$5512.40$$

3. (a) $j = 11.5\% = 0.115$

$f = e^j - 1 = e^{0.115} - 1 = 1.1218734 - 1 = 0.1218734 = 12.187\%$

(b) $j = 7.3\% = 0.073$

$f = e^{0.073} - 1 = 1.0757305 - 1 = 0.0757305 = 7.573\%$

B. 1. $FV_1 = 1000.00(1.02)^{20} = 1000.00(1.4859474) = \$1485.95$

$FV_2 = 1000.00e^{(0.08)(5)} = 1000.00e^{0.4} = 1000.00(1.4918247) = \$1491.82$

Difference in interest $= 1491.82 - 1485.95 = \$5.87$

3. Let the focal date be today; $j = 0.08$

Let the single payment be $\$x$.

$x = 1600.00e^{-(0.08)\left(\frac{15}{12}\right)} + 1600.00e^{-(0.08)\left(\frac{30}{12}\right)}$

$= 1600.00\left(e^{-0.1} + e^{-0.2}\right)$

$= 1600.00(0.9048374 + 0.8187308)$

$= 1600.00(1.7235682)$

$= \$2757.71$

5. The value of $\$1$ in one year at 5% compounded continuously.

$FV_1 = e^{0.05} = 1.0512711$

The value of $\$1$ in one year at $i\%$ per quarter, $FV_2 = (1 + i)^4$

For $FV_1 = FV_2$ $(1 + i)^4 = 1.0512711$

$(1 + i) = 1.0512711^{\frac{1}{4}}$

$1 + i = 1.0125785$

$i = 1.25785\%$

The nominal annual rate is $1.25785\% \times 4 = 5.031\%$.

7. (a) Let $PV = \$1$; let $FV = \$2$; $j = 6.0\% = 0.06$

$FV = PVe^{nj}$

$2 = e^{0.06n}$

$\ln 2 = 0.06n \ln e$

$0.6931472 = 0.06n$

$n = 11.552453$

At 6% compounded continuously, money doubles in 11 years 202 days.

(b) Let PV = $1; let FV = $2; $j = 11.4\% = 0.114$

$$2 = e^{0.114n}$$
$$\ln 2 = 0.114n \ln e$$
$$0.6931472 = 0.114n$$
$$n = 6.0802386$$

At 11.4% compounded continuously, money doubles in 6 years 30 days.

9. Amount of $1 in one year at 7% compounded quarterly.

$$FV_1 = 1.0175^4 = 1.0718590$$

Amount of $1 in one year at $j\%$ compounded continuously, $FV_2 = e^j$

For $FV_1 = FV_2$, $e^j = 1.0718590$
$$j \ln e = \ln 1.0718590$$
$$j = 0.0693945$$

The nominal rate compounded continuously is 6.94%.

Review Exercise

1. $PV = 6000.00$; $i = 0.03$; $n = 2\left(6\dfrac{7}{12}\right) = 13.1666667$

$$FV = 6000.00(1.03)^{13.1666667}$$
$$= 6000.00(1.4757862)$$
$$= 8854.72$$
$$I = 8854.72 - 6000.00 = \$2854.72$$

3. Interest period: 1995-11-15 to 2005-06-15
         9 years 7 months

$PV = 5000.00$; $i = 0.02$; $n = 4\left(9\dfrac{7}{12}\right) = 38.3333333$

Maturity value, $FV = 5000.00(1.02)^{38.3333333}$
$$= 5000.00(2.1363542)$$
$$= \$10\,681.77$$

5. PV = 16 500.00; $i = 0.0125$; $n = 180$

    Maturity value $= 16\,500.00(1.0125)^{180}$

    $\qquad\qquad\quad = 16\,500.00(9.3563345)$

    $\qquad\qquad\quad = 154\,379.52$

    FV $= 154\,379.52$; $i = 0.045$; $n = 2\left(15 - 3\dfrac{4}{12}\right) = 23.3333333$

    Proceeds $= 154\,379.52(1.045)^{-23.3333333}$

    $\qquad\qquad = 154\,379.52(0.3580579)$

    $\qquad\qquad = \$55\,276.80$

7. PV = 2000.00; $i = 0.0225$; $n = 2\left(3\dfrac{4}{12}\right) = 6.6666667$

    FV $= 2000.00(1.0225)^{6.6666667}$

    $\quad\ = 2000.00(1.1599042)$

    $\quad\ = 2319.81$

    $\ \ I = 2319.81 - 2000.00 = \$319.81$

9. FV = 1600.00; $i = 0.095$; $n = 2\dfrac{8}{12} = 2.6666667$

    Proceeds $= 1600.00(1.095)^{-2.6666667}$

    $\qquad\qquad = 1600.00(0.7850470)$

    $\qquad\qquad = \$1256.08$

11. Value of \$1 in one year at $i\%$ monthly, $FV_1 = (1 + i)^{12}$

    Value of \$1 in one year at 6.2% effective, $FV_2 = 1.062$

    For $FV_1 = FV_2$, $(1 + i)^{12} = 1.062$

    $$1 + i = 1.062^{\frac{1}{12}}$$
    $$1 + i = 1.0050254$$
    $$i = 0.50254\%$$

    The nominal rate is $0.50254\% \times 12 = 6.03\%$.

13. Let PV = \$1; FV = \$3; $i = 5.0\% = 0.05$; $m = 2$

    $$3 = 1(1.05)^n$$
    $$\ln 3 = n\ln 1.05$$
    $$1.0986123 = 0.0487902n$$
    $$n = 22.517069\ (\text{half - years})$$
    $$= 11.2585345\ (\text{years})$$
    $$= 11\ \text{years}\ 95\ \text{days}$$

15. (a) $f = 1.00375^{12} - 1 = 1.0459398 - 1 = 0.0459398 = 4.59\%$

 (b) $3000.00 = 2000.00(1 + i)^{28}$; $m = 4$

 $(1 + i)^{28} = 1.5$

 $1 + i = 1.5^{\frac{1}{28}}$

 $1 + i = 1.01458625$

 $f = 1.01458625^4 - 1 = 1.0596340 - 1 = 0.059634 = 5.96\%$

17. (a) $9517.39 = 7500.00(1.015)^n$; $m = 2$

 $1.015^n = 1.2689853$

 $n \ln 1.015 = \ln 1.2689853$

 $0.0148886n = 0.2382176$

 $n = 16$ (half - years)

 It will take $\dfrac{16}{2} = 8$ years.

 (b) Let $PV = \$1$; $FV = \$3$; $i = 2.25\% = 0.0225$; $m = 4$

 $3 = 1(1.0225)^n$

 $\ln 3 = n \ln 1.0225$

 $1.0986123 = 0.0222506n$

 $n = 49.3745023$ (quarters)

 $= 12.3436256$ (years)

 Money will triple in 12 years 126 days.

 (c) $PV = 10\,000.00$; $FV = 13\,684.00$; $i = 0.875\% = 0.00875$; $m = 12$

 $13\,684.00 = 10\,000.00(1.00875)^n$

 $1.00875^n = 1.3684$

 $n \ln 1.00875 = \ln 1.3684$

 $0.0087119n = 0.3136422$

 $n = 36.001584$ (months)

 $= 3$ (years)

 It will take 3 years.

19. Let the focal date be now; $i = 5.5\% = 0.055$; $m = 2$

$$5000.00(1.055)^{-1} + 6000.00(1.055)^{-2.5} = 5000.00 + 6000.00(1.055)^{-n}$$
$$5000.00(0.9478673) + 6000.00(0.8747196) = 5000.00 + 6000.00(1.055)^{-n}$$
$$4739.34 + 5248.32 = 5000.00 + 6000.00(1.055)^{-n}$$
$$4987.66 = 6000.00(1.055)^{-n}$$
$$1.055^{-n} = 0.8312767$$
$$-n\ln 1.055 = \ln 0.8312767$$
$$-0.0535408n = -0.1847926$$
$$n = 3.4514352 \text{ (half-years)}$$
$$= 1.7257176 \text{ (years)}$$

The payment of $6000.00 should be made 1 year 265 days from now.

21. $PV = 3300.00$; $i = 0.625\% = 0.00625$; $n = 72$

Maturity value $= 3300.00(1.00625)^{72} = 3300.00(1.5661174) = \$5168.19$

$FV = 5168.19$; $i = 0.035$; $n = 2\left(6 - 3\dfrac{5}{12}\right) = 5.1666667$

$$\begin{aligned} \text{Proceeds} &= 5168.19(1.035)^{-5.1666667} \\ &= 5168.19(0.8371595) \\ &= \$4326.60 \end{aligned}$$

23. $PV = 700.00$; $n = 7$; $j = 8.5\% = 0.085$

$$FV = 700.00e^{7(0.085)} = 700.00e^{0.595} = 700.00(1.8130309) = \$1269.12$$

25. Value of $1 in one year:

at effective rate f, $FV_1 = 1 + f$

at 6.75% compounded continuously, $FV_2 = e^{0.0675}$

For $FV_1 = FV_2$, $1 + f = e^{0.0675}$
$$1 + f = 1.0698303$$
$$f = 0.0698303$$
$$= 6.983\%$$

27. (a) $PV = 10\,000.00e^{-(0.04)(10)} = 10\,000.00e^{-0.4} = 10\,000.00(0.6703200) = \$6703.20$

(b) $PV = 7000.00e^{-(0.055)\left(4\frac{4}{12}\right)} = 7000.00e^{-0.2383333} = 7000.00(0.7879400) = \$5515.58$

29. Let the focal date be now; $j = 7.5\% = 0.075$

$$2000.00e^{-0.075} + 4000.00e^{-5(0.075)} = 3000.00e^{-2(0.075)} + 4500.00e^{-0.075n}$$

$$2000.00(0.9277435) + 4000.00(0.6872893) = 3000.00(0.8607080) + 4500.00e^{-0.075n}$$

$$1855.49 + 2749.16 = 2582.12 + 4500.00e^{-0.075n}$$

$$2022.53 = 4500.00e^{-0.075n}$$

$$e^{-0.075n} = 0.4494511$$

$$-0.075n \ln e = \ln 0.4494511$$

$$-0.075n = -0.7997282$$

$$n = 10.6630427 \text{ (years)}$$

The date of the second payment is 10 years 242 days from now.

Self-Test

1. Maturity value: $PV = 9200.00$; $i = 0.5\%$; $n = 120$

$FV = 9200.00(1.005)^{120} = 9200.00(1.8193967) = \$16\,738.45$

Proceeds: $PV = 12\,915.60$; $FV = 16\,738.45$; $i = 2.5\%$

$$16\,738.45 = 12\,915.60(1.025)^n$$

$$1.025^n = 1.2959870$$

$$n \ln 1.025 = \ln 1.2959870$$

$$0.0246926n = 0.2592726$$

$$n = 10.50 \text{ (half-years)}$$

The date of discount is 63 months before the due date.

3. $i = 0.45\%$; $m = 12$

$$f = 1.0045^{12} - 1 = 1.05535675 - 1 = 5.535675\%$$

5. $PV = 6900.00$; $FV = 6900.00 + 3000.00 = 9900.00$; $n = 10$; $m = 2$

$$9900.00 = 6900.00(1 + i)^{10}$$

$$(1 + i)^{10} = 1.4347826$$

$$1 + i = 1.4347826^{0.10}$$

$$1 + i = 1.0367609$$

$$i = 3.67609\%$$

The nominal annual rate is $3.676\% \times 2 = 7.352\%$

7. $f = 10.25\%$; $m = 2$

$$1.1025 = (1 + i)^2$$

$$1 + i = 1.1025^{0.5}$$

$$1 + i = 1.05$$

$$i = 0.05$$

The nominal annual rate is 10.0%.

9.   FV = 1100.00; i = 7.5%; $n = 3\frac{7}{12} = 3.5833333$

PV = $1100.00(1.075)^{-3.5833333} = 1100.00(0.7717080) = \$848.88$

11. i = 2%; m = 4

$1 + f = 1.02^4 = 1.0824322$

$e^j = 1.0824322$

$j \ln e = \ln 1.0824322$

$j = 0.0792105$

$j = 7.921\%$

11 Ordinary simple annuities

Exercise 11.1

A. 1. (a) annuity certain   (b) annuity due   (c) general annuity

3. (a) perpetuity   (b) deferred annuity   (c) general annuity

5. (a) annuity certain   (b) deferred annuity due   (c) simple annuity

Exercise 11.2

A. 1. PMT = 1500.00; $i = 1.25\% = 0.0125$; $n = 7.5 \times 4 = 30$

$$FV_n = 1500.00\left[\frac{1.0125^{30} - 1}{0.0125}\right] = 1500.00(36.1290688) = \$54\,193.60$$

3. PMT = 700.00; $i = 3.5\% = 0.035$; $n = 40$

$$FV_n = 700.00\left[\frac{1.035^{40} - 1}{0.035}\right] = 700.00(84.5502778) = \$59\,185.19$$

5. PMT = 320.00; $i = 2.6\% = 0.026$; $n = 8 \times 4 + \dfrac{9}{12} \times 4 = 35$

$$FV_n = 320.00\left[\frac{1.026^{35} - 1}{0.026}\right] = 320.00(55.9846160) = \$17\,915.08$$

B. 1. PMT = 200.00; $i = 1.25\%$; $n = 48$

$$FV_n = 200.00\left[\frac{1.0125^{48} - 1}{0.0125}\right] = 200.00(65.2283882) = \$13\,045.68$$

3. PMT = 1500.00; $i = 3.5\%$; $n = 30$

$$FV_n = 1500.00\left[\frac{1.035^{30} - 1}{0.035}\right] = 1500.00(51.6226773) = \$77\,434.02$$

Interest = $77\,434.02 - 1500.00(30) = 77\,434.02 - 45\,000.00 = \$32\,434.02$

5. (a) After 15 years: PMT = 25.00; i = 0.3%; n = 180

$$FV_n = 25.00\left[\frac{1.003^{180} - 1}{0.003}\right] = 25.00(238.2067436) = \$5955.17$$

Amount 25 years from now:
PV = 5955.17; i = 0.3%; n = 10 × 12 = 120

$$FV = 5955.17(1.003)^{120} = 5955.17(1.4325572) = \$8531.12$$

(b) Contribution = 25.00(180) = \$4500.00

(c) Interest = 8531.12 − 4500.00 = \$4031.12

7. After 7 years: PMT = 1375.00; i = 1.75%; n = 28

$$FV_n = 1375.00\left[\frac{1.0175^{28} - 1}{0.0175}\right] = 1375.00(35.7378798) = \$49\,139.58$$

3 years later:
PV = 49 139.58; i = 4%; n = 6

$$FV = 49\,139.58(1.04)^{6} = 49\,139.58(1.2653190) = \$62\,177.25$$

Exercise 11.3

A. 1. PMT = 1600.00; i = 4.25%; n = 7

$$PV_n = 1600.00\left[\frac{1 - (1.0425)^{-7}}{0.0425}\right] = 1600.00(5.9469928) = \$9515.19$$

3. PMT = 4000.00; i = 7.5%; n = 12

$$PV_n = 4000.00\left[\frac{1 - 1.075^{-12}}{0.075}\right] = 4000.00(7.7352783) = \$30\,941.11$$

5. PMT = 250.00; i = 1.1%; n = 57

$$PV_n = 250.00\left[\frac{1 - 1.011^{-57}}{0.011}\right] = 250.00(42.1796480) = \$10\,544.91$$

B. 1. PMT = 375.00; $i = 3.5\%$; $n = 30$

$$PV_n = 375.00\left[\frac{1 - 1.035^{-30}}{0.035}\right] = 375.00(18.3920454) = \$6897.02$$

3. (a) PMT = 600.00; $i = 1.9\%$; $n = 20$

$$PV_n = 600.00\left[\frac{1 - 1.019^{-20}}{0.019}\right] = 600.00(16.5103333) = \$9906.20$$

(b) Interest = $600.00(20) - 9906.20 = 12\,000.00 - 9906.20 = \$2093.80$

5. (a) PMT = 69.33; $i = 0.9\%$; $n = 36$

$$PV_n = 69.33\left[\frac{1 - 1.009^{-36}}{0.009}\right] = 69.33(30.6334204) = \$2123.82$$

Cash price = $400.00 + 2123.82 = \$2523.82$

(b) Cost of financing = $36(69.33) - 2123.82 = 2495.88 - 2123.82 = \$372.06$

7. Investment at date of retirement:
PMT = 450.00; $i = 1.25\%$; $n = 60$

$$PV_n = 450.00\left[\frac{1 - 1.0125^{-60}}{0.0125}\right] = 450.00(42.0345918) = \$18915.57$$

Investment 8 years earlier:
FV = 18 915.57; $i = 1.25\%$; $n = 32$
PV = $18\,915.57(1.0125)^{-32} = 18\,915.57(0.6719841) = \$12\,710.96$

9. (a) PMT = 234.60; $i = 0.6\%$; $n = 72$

$$PV_n = 234.60\left[\frac{1 - 1.006^{-72}}{0.006}\right] = 234.60(58.3253429) = \$13\,683.13$$

(b) PMT = 234.60; $i = 0.6\%$; $n = 6$

$$FV_n = 234.60\left[\frac{1.006^6 - 1}{0.006}\right] = 234.60(6.0907232) = \$1428.88$$

177

    (c)   Payout amount   = 1428.89 + Present value of the remaining payments

$$= 1428.89 + 234.60\left[\frac{1 - 1.006^{-66}}{0.006}\right]$$

$$= 1428.89 + 234.60(54.3660808)$$
$$= 1428.89 + 12\,754.28$$
$$= 14\,183.17$$

    (d)   Interest $= 14\,183.17 - 13\,683.13 = \$500.04$

    (e)   Payment required after six months $1428.88
          Normal payment $= 6 \times 234.60$ 1407.60
          Additional interest $ 21.28

Exercise 11.4

A.   1.   PMT $= 2500.00$; $i = 2\%$; $n = 14$; $c = 2$

$$p = (1 + i)^c - 1 = 1.02^2 - 1 = 1.0404 - 1 = 0.0404 = 4.04\%$$

$$FV_{nc} = 2500.00\left[\frac{1.0404^{14} - 1}{0.0404}\right] = 2500.00(18.3421833) = \$45\,855.46$$

     3.   PMT $= 72.00$; $i = 1.5\%$; $n = 180$; $c = \dfrac{2}{12} = \dfrac{1}{6}$

$$p = (1.015)^{\frac{1}{6}} - 1 = 1.0024845 - 1 = 0.0024845 = 0.24845\%$$

$$FV_{nc} = 72.00\left[\frac{1.0024845^{180} - 1}{0.0024845}\right] = 72.00(226.6353498) = \$16\,317.75$$

     5.   PMT $= 1750.00$; $i = 3.5\%$; $n = 24$; $c = 1$

$$FV_n = 1750.00\left[\frac{1.035^{24} - 1}{0.035}\right] = 1750.00(36.6665282) = \$64\,166.42$$

     7.   PMT $= 7500.00$; $i = 1.5\%$; $n = 4$; $c = 4$

$$p = 1.015^4 - 1 = 1.0613636 - 1 = 0.0613636 = 6.13636\%$$

$$FV_{nc} = 7500.00\left[\frac{1.0613636^4 - 1}{0.0613636}\right] = 7500.00(4.3834746) = \$32\,876.06$$

B. 1. PMT $= 425.00$; $i = 0.75\%$; $n = 36$; $c = \dfrac{12}{4} = 3$

$p = 1.0075^3 - 1 = 1.0226692 - 1 = 2.26692\%$

$FV_{nc} = 425.00\left[\dfrac{1.0226692^{36} - 1}{0.0226692}\right] = 425.00(54.7494570) = \$23\,268.52$

3. PMT $= 15.00$; $i = 1.25\%$; $n = 120$; $c = \dfrac{4}{12} = \dfrac{1}{3}$

$p = 1.0125^{\frac{1}{3}} - 1 = 1.0041494 - 1 = 0.41494\%$

$FV_{nc} = 15.00\left[\dfrac{1.0041494^{120} - 1}{0.0041494}\right] = 15.00(155.1102638) = \$2326.65$

5. (a) PMT $= 1000.00$; $i = 1.5\%$; $n = 10$; $c = \dfrac{4}{1} = 4$

$p = 1.015^4 - 1 = 1.0613636 - 1 = 6.13636\%$

$FV_{nc} = 1000.00\left[\dfrac{1.0613636^{10} - 1}{0.0613636}\right] = 1000.00(13.2655068) = \$13\,265.51$

(b) Interest $= 13\,265.51 - 10(1000.00) = 13\,265.51 - 10\,000.00 = \$3265.51$

7. Amount after 10 years:

PMT $= 500.00$; $i = 2.25\%$; $n = 40$; $c = \dfrac{2}{4} = \dfrac{1}{2}$

$p = 1.0225^{\frac{1}{2}} - 1 = 1.0111874 - 1 = 1.11874\%$

$FV_{nc} = 500.00\left[\dfrac{1.0111874^{40} - 1}{0.0111874}\right] = 500.00(50.1017141) = \$25\,050.86$

Amount 5 years later:
PV $= 25\,050.86$; $i = 2.25\%$; $n = 10$

$FV = 25\,050.86\left(1.0225^{10}\right) = 25\,050.86(1.2492034) = \$31\,293.62$

CHAPTER 11

Exercise **11.5**

A. 1. $\text{PMT} = 1400.00$; $i = 0.5\%$; $n = 48$; $c = \dfrac{12}{4} = 3$

$p = 1.005^3 - 1 = 1.0150751 - 1 = 0.0150751 = 1.50751\%$

$\text{PV}_{nc} = \text{PMT}\left[\dfrac{1 - (1 + p)^{-n}}{p}\right]$

$= 1400.00\left[\dfrac{1 - 1.0150751^{-48}}{0.0150751}\right] = 1400.00(33.9880425) = \$47\,583.23$

3. $\text{PMT} = 3000.00$; $i = 6\%$; $n = 16$; $c = \dfrac{1}{4}$

$p = 1.06^{\frac{1}{4}} - 1 = 1.0146738 - 1 = 0.0146738 = 1.46738\%$

$\text{PV}_{nc} = 3000.00\left[\dfrac{1 - 1.0146738^{-16}}{0.0146738}\right] = 3000.00(14.1685017) = \$42\,505.51$

5. $\text{PMT} = 95.00$; $i = 0.375\%$; $n = 60$; $c = 1$

$\text{PV}_{n} = 95.00\left[\dfrac{1 - 1.00375^{-60}}{0.00375}\right] = 95.00(53.6393804) = \$5095.74$

7. $\text{PMT} = 1890.00$; $i = 1.75\%$; $n = 30$; $c = \dfrac{4}{2} = 2$

$p = 1.0175^2 - 1 = 1.0353063 - 1 = 0.0353063 = 3.53063\%$

$\text{PV}_{nc} = 1890.00\left[\dfrac{1 - 1.0353063^{-30}}{0.0353063}\right] = 1890.00(18.3216666) = \$34\,627.95$

B. 1. $\text{PMT} = 250.00$; $i = 0.25\%$; $n = 48$; $c = \dfrac{12}{4} = 3$

$p = 1.0025^3 - 1 = 1.0075188 - 1 = 0.75188\%$

$\text{PV}_{nc} = 250.00\left[\dfrac{1 - 1.0075188^{-48}}{0.0075188}\right] = 250.00(40.1674888) = \$10\,041.87$

3. $\text{PMT} = 825.00$; $i = 3.5\%$; $n = 64$; $c = \dfrac{2}{4} = \dfrac{1}{2}$

   $p = 1.035^{\frac{1}{2}} - 1 = 1.0173495 - 1 = 1.73495\%$

   $\text{PV}_{nc} = 825.00\left[\dfrac{1 - 1.0173495^{-64}}{0.0173495}\right] = 825.00(38.4685636) = \$31\,736.56$

5. (a) $\text{PMT} = 2500.00$; $i = 0.5\%$; $n = 12$; $c = \dfrac{12}{2} = 6$

   $p = 1.005^{6} - 1 = 1.0303775 - 1 = 3.03775\%$

   $\text{PV}_{nc} = 2500.00\left[\dfrac{1 - 1.0303775^{-12}}{0.0303775}\right] = 2500.00(9.9316103) = \$24\,829.03$

   Purchase price $= 24\,829.03 + 5000.00 = \$29\,829.03$

   (b) Interest $= 12(2500.00) - 24\,829.03 = 30\,000.00 - 24\,829.03 = \$5170.97$

7. $\text{PMT} = 715.59$; $i = 5\%$; $n = 300$; $c = \dfrac{2}{12} = \dfrac{1}{6}$

   $p = 1.05^{\frac{1}{6}} - 1 = 1.0081648 - 1 = 0.81648\%$

   $\text{PV}_{nc} = 715.59\left[\dfrac{1 - 1.0081648^{-300}}{0.0081648}\right] = 715.59(111.7963793) = \$80\,000.37$

9. Amount at retirement:

   $\text{PMT} = 1200.00$; $i = 0.55\%$; $n = 80$; $c = \dfrac{12}{4} = 3$

   $p = 1.0055^{3} - 1 = 1.0165909 - 1 = 1.65909\%$

   $\text{PV}_{nc} = 1200.00\left[\dfrac{1 - 1.0165909^{-80}}{0.0165909}\right] = 1200.00(44.114330) = \$52\,937.20$

   Investment 7 years earlier:
   $\text{FV} = 52\,937.20$; $i = 0.55\%$; $n = 84$
   $\text{PV} = 52\,937.20\left(1.0055^{-84}\right) = 52\,937.20(0.6308204) = \$33\,393.87$

1. (a) PMT = 360.00; i = 1.75%; n = 48

$$FV_n = 360.00\left[\frac{1.0175^{48} - 1}{0.0175}\right] = 360.00(74.2627843) = \$26\,734.60$$

(b) Amount deposited = 48(360.00) = \$17\,280.00

(c) Interest = 26\,734.60 − 17\,280.00 = \$9454.60

3. PMT = 75.90; i = 0.75%; n = 48

$$FV_n = 75.90\left[\frac{1.0075^{48} - 1}{0.0075}\right] = 75.90(57.5207111) = \$4365.82$$

Interest = 4365.82 − 48(75.90) = 4365.82 − 3643.20 = \$722.62

5. PMT = 375.00; i = 0.25%; n = 32; $c = \dfrac{12}{4} = 3$

$p = 1.0025^3 - 1 = 1.0075188 - 1 = 0.75188\%$

$$FV_{nc} = 375.00\left[\frac{1.0075188^{32} - 1}{0.0075188}\right] = 375.00(36.0256762) = \$13\,509.63$$

7. PMT = 3500.00; i = 1.75%; n = 30; $c = \dfrac{4}{2} = 2$

$p = 1.0175^2 - 1 = 1.0353063 - 1 = 3.53063\%$

$$PV_{nc} = 3500.00\left[\frac{1 - 1.0353063^{-30}}{0.0353063}\right] = 3500.00(18.3216666) = \$64\,125.83$$

9. Amount after 12 years:
    PMT = 1500.00; i = 4.0%; n = 24

$$FV_n = 1500.00\left[\frac{1.04^{24} - 1}{0.04}\right] = 1500.00(39.0826041) = \$58\,623.91$$

Amount 7 years later:
FV = 58\,623.91(1.04)^{14} = 58\,623.91(1.7316764) = \$101\,517.64

11. $PMT = 500.00;\ i = 3.0\%;\ n = 80;\ c = \dfrac{2}{4} = \dfrac{1}{2}$

$p = 1.03^{\frac{1}{2}} - 1 = 1.0148892 - 1 = 1.48892\%$

$FV_{nc} = 500.00\left[\dfrac{1.0148892^{80} - 1}{0.0148892}\right] = 500.00(151.9254879) = \$75\,962.74$

13.   Amount after 4 years:
$PMT = 250.00;\ i = 1\%;\ n = 16$

$FV_n = 250.00\left[\dfrac{1.01^{16} - 1}{0.01}\right] = 250.00(17.2578645) = \$4314.47$

Amount 9 years later:
$PV = 4314.47;\ i = 1.25\%;\ n = 36$
$FV = 4314.47(1.0125)^{36} = 4314.47(1.5639438) = \$6747.59$

Amount after 10 years:
$PMT = 250.00;\ i = 1.25\%;\ n = 24$

$FV_n = 250.00\left[\dfrac{1.0125^{24} - 1}{0.0125}\right] = 250.00(27.7880840) = \$6947.02$

Amount 3 years later:
$PV = 6947.02;\ i = 1.25\%;\ n = 12$
$FV = 6947.02(1.0125)^{12} = 6947.02(1.1607545) = \$8063.78$
RRSP balance $= 6747.59 + 8063.78 = \$14\,811.37$

15.   (a) $PMT = 750.00;\ i = 2.25\%;\ n = 32$

$PV_n = 750.00\left[\dfrac{1 - 1.0225^{-32}}{0.0225}\right] = 750.00(22.6376742) = \$16\,978.26$

(b) $PMT = 750.00;\ i = 2.25\%;\ n = 4$

$FV_n = 750.00\left[\dfrac{1.0225^{4} - 1}{0.0225}\right] = 750.00(4.1370364) = \$3102.78$

(c)   Present value of remaining payments:
$PMT = 750.00;\ i = 2.25\%;\ n = 28$

$PV_n = 750.00\left[\dfrac{1 - 1.0225^{-28}}{0.0225}\right] = 750.00(20.6078276) = \$15\,455.87$

Payout figure $= 3102.78 + 15\,455.87 = \$18\,558.65$

(d)   Missed payment replacement figure $3102.78
      Normal payment $= 4 \times 750.00$ 3000.00
      Additional interest $ 102.78

1. $PMT = 2400.00$; $i = 1.375\%$; $n = 32$

$$PV_n = 2400.00\left[\frac{1 - 1.01375^{-32}}{0.01375}\right] = 2400.00(25.7476472) = \$61\,794.35$$

$FV = 61\,794.35$; $i = 1.375\%$; $n = 40$

$PV = 61\,794.35(1.01375)^{-40} = 61\,794.35(0.5791157) = \$35\,786.08$

3. $PMT = 480.00$; $i = 2.25\%$; $n = 96$; $c = \dfrac{2}{12} = \dfrac{1}{6} = 0.1666667$

$p = 1.0225^{0.1666667} - 1 = 1.0037153 - 1 = 0.37153\%$

$$FV_{nc} = 480.00\left[\frac{1.0037153^{96} - 1}{0.0037153}\right] = 480.00(115.0967041) = \$55\,246.42$$

$PV = 55\,246.42$; $i = 2.25\%$; $n = 10$

$FV = 55\,246.42(1.0225)^{10} = 55\,246.42(1.2492034) = \$69\,014.02$

5. Balance after 2 years: $PV = 7500.00$; $i = 2.625\%$; $n = 8$

$FV_2 = 7500.00(1.02625)^8 = 7500.00(1.2303406) = \$9227.55$

Balance after 5 years: $PV = 9227.55 - 4500.00 = 4727.55$; $n = 12$

$FV_5 = 4727.55(1.02625)^{12} = 4727.55(1.3647027) = \$6451.70$

Balance after 9 years: $PV = 6451.70 - 3500.00 = 2951.70$; $n = 16$

$FV_9 = 2951.70(1.02625)^{16} = 2951.70(1.5137380) = \$4468.10$

7. Accumulated value of the deposits for first 5 years:
$PMT = 540.00$; $i = 1.25\%$; $n = 20$

$$FV_n = 540.00\left[\frac{1.0125^{20} - 1}{0.0125}\right] = 540.00(22.5629785) = \$12\,184.01$$

$PV = 12\,184.01$; $i = 1.375\%$; $n = 32$

$FV = 12\,184.01(1.01375)^{32} = 12\,184.01(1.5480599) = \$18\,861.58$

Accumulated value of the deposits for remaining 8 years:
$PMT = 540.00$; $i = 1.375\%$; $n = 32$

$$FV_n = 540.00\left[\frac{1.01375^{32} - 1}{0.01375}\right] = 540.00(39.8588992) = \$21\,523.81$$

Balance $= 18\,861.58 + 21\,523.81 = \$40\,385.39$

12 Other simple annuities

A. 1. $\text{FV}_n(\text{due}) = 3000.00(1.02)\left[\dfrac{1.02^{32} - 1}{0.02}\right]$

$= 3000.00(1.02)(44.2270296)$

$= \$135\,334.71$

$\text{PV}_n(\text{due}) = 3000.00(1.02)\left[\dfrac{1 - 1.02^{-32}}{0.02}\right]$

$= 3000.00(1.02)(23.4683348)$

$= \$71\,813.10$

3. $\text{FV}_n(\text{due}) = 2000.00(1.028)\left[\dfrac{1.028^{24} - 1}{0.028}\right]$

$= 2000.00(1.028)(33.5766970)$

$= \$69\,033.69$

$\text{PV}_n(\text{due}) = 2000.00(1.028)\left[\dfrac{1 - 1.028^{-24}}{0.028}\right]$

$= 2000.00(1.028)(17.3062598)$

$= \$35\,581.67$

5. $\text{FV}_n(\text{due}) = 65.00(1.0075)\left[\dfrac{1.0075^{240} - 1}{0.0075}\right]$

$= 65.00(1.0075)(667.886870)$

$= \$43\,738.24$

$\text{PV}_n(\text{due}) = 65.00(1.0075)\left[\dfrac{1 - 1.0075^{-240}}{0.0075}\right]$

$= 65.00(1.0075)(111.1449540)$

$= \$7278.61$

B. 1. $\text{PMT} = 300.00; \ i = 0.5\%; \ n = 84$

$\text{FV}_n(\text{due}) = 300.00(1.005)\left[\dfrac{1.005^{84} - 1}{0.005}\right] = 300.00(1.005)(104.0739272) = \$31\,378.29$

3. PMT = 2500.00; $i = 4.75\%$; $n = 20$

$$PV_n(\text{due}) = 2500.00(1.0475)\left[\frac{1 - 1.0475^{-20}}{0.0475}\right]$$

$$= 2500.00(1.0475)(12.7306690) = \$33\,338.44$$

5. (a) PMT = 52.50; $i = 1.75\%$; $n = 30$

$$PV_n(\text{due}) = 52.50(1.0175)\left[\frac{1 - 1.0175^{-30}}{0.0175}\right] = 52.50(1.0175)(23.1858493) = \$1238.56$$

   (b) Paid in instalments = $30(52.50) = \$1575.00$

   (c) Interest = $1575.00 - 1238.56 = \$336.44$

7. (a) PMT = 300.00; $i = 1.25\%$; $n = 64$

$$FV_n(\text{due}) = 300.00(1.0125)\left[\frac{1.0125^{64} - 1}{0.0125}\right]$$

$$= 300.00(1.0125)(97.1625929) = \$29\,513.14$$

   (b) Total deposits = $64(300.00) = \$19\,200.00$
       Interest = $29\,513.14 - 19\,200.00 = \$10\,313.14$

9. PMT = 64.00; $i = 0.4\%$; $n = 36$

$$PV_n(\text{due}) = 64.00(1.004)\left[\frac{1 - 1.004^{-36}}{0.004}\right] = 64.00(1.004)(33.4658759) = \$2150.38$$

Exercise 12.2

A. 1. PMT = 1500.00; $i = 1.25\%$; $n = 20$; $c = \dfrac{4}{2} = 2$

$$p = 1.0125^2 - 1 = 1.02515625 - 1 = 2.515625\%$$

$$FV_{nc}(\text{due}) = 1500.00(1.02515625)\left[\frac{1.02515625^{20} - 1}{0.02515625}\right]$$

$$= 1500.00(1.02515625)(25.5848731)$$

$$= \$39\,342.74$$

3. PMT = 650.00; i = 1%; n = 24; $c = \dfrac{12}{4} = 3$

$p = 1.01^3 - 1 = 1.030301 - 1 = 3.0301\%$

$PV_{nc}(\text{due}) = 650.00(1.030301)\left[\dfrac{1 - 1.030301^{-24}}{0.030301}\right]$

$\quad = 650.00(1.030301)(16.8807602)$

$\quad = \$11\,304.97$

B. 1. PMT = 5000.00; i = 1.5%; n = 10; $c = \dfrac{4}{2} = 2$

$p = 1.015^2 - 1 = 1.030225 - 1 = 3.0225\%$

$FV_{nc}(\text{due}) = 5000.00(1.030225)\left[\dfrac{1.030225^{10} - 1}{0.030225}\right]$

$\quad = 5000.00(1.030225)(11.4757653)$

$\quad = \$59\,113.10$

3. PMT = 750.00; i = 2%; n = 36; $c = \dfrac{4}{12} = \dfrac{1}{3}$

$p = 1.02^{\frac{1}{3}} - 1 = 1.0066227 - 1 = 0.66227\%$

$PV_{nc}(\text{due}) = 750.00(1.0066227)\left[\dfrac{1 - 1.0066227^{-36}}{0.0066227}\right]$

$\quad = 750.00(1.0066227)(31.9366051)$

$\quad = \$24\,111.08$

Exercise 12.3

A. 1. PMT(due) = 850.00; i = 7.5%; n = 10; d = 3

$PV_n = 850.00(1.075)\left[\dfrac{1 - 1.075^{-10}}{0.075}\right] = 850.00(1.075)(6.8640810) = 6272.05$

$PV_n(\text{defer}) = 6272.05\left(1.075^{-3}\right) = 6272.05(0.8049606) = \$5048.75$

3. PMT = 125.00; i = 3.5%; n = 30; d = 16

$PV_n = 125.00\left[\dfrac{1 - 1.035^{-30}}{0.035}\right] = 125.00(18.3920454) = 2299.01$

$PV_n(\text{defer}) = 2299.01(1.035)^{-16} = 2299.01(0.5767059) = \$1325.85$

5. PMT(due) = 85.00; i = 0.5%; n = 180; d = 240

$$PV_n = 85.00(1.005)\left[\frac{1 - 1.005^{-180}}{0.005}\right] = 85.00(1.005)(118.5035147) = 10\,123.16$$

$$PV_n(\text{defer}) = 10\,123.16\left(1.005^{-240}\right) = 10\,123.16(0.3020961) = \$3058.17$$

7. PMT = 720.00; i = 1%; n = 40; d = 16; $c = \dfrac{12}{4} = 3$

$$p = 1.01^3 - 1 = 1.030301 - 1 = 0.030301 = 3.0301\%$$

$$PV_{nc} = 720.00\left[\frac{1 - 1.030301^{-40}}{0.030301}\right] = 720.00(23.0027135) = 16\,561.95$$

$$PV_{nc}(\text{defer}) = 16\,561.95\left(1.030301^{-16}\right) = 16\,561.95(0.6202604) = \$10\,272.72$$

9. PMT(due) = 225.00; i = 9%; n = 32; d = 24; $c = \dfrac{1}{4}$

$$p = 1.09^{\frac{1}{4}} - 1 = 1.0217782 - 1 = 0.0217782 = 2.17782\%$$

$$PV_{nc}(\text{due}) = 225.00(1.0217782)\left[\frac{1 - 1.0217782^{-32}}{0.0217782}\right] = 225.00(1.0217782)(22.8730575)$$

$$= 5258.52$$

$$PV_{nc}(\text{defer}) = 5258.52\left(1.0217782^{-24}\right) = 5258.52(0.5962671) = \$3135.48$$

B. 1. PMT = 5000.00; i = 2%; n = 40; d = 8

$$PV_n = 5000.00\left[\frac{1 - 1.02^{-40}}{0.02}\right] = 5000.00(27.3554792) = 136\,777.40$$

$$PV_n(\text{defer}) = 136\,777.40\left(1.02^{-8}\right) = 136\,777.40(0.8534904) = \$116\,738.20$$

Purchase price = $10\,000.00 + 116\,738.20 = \$126\,738.20$

3. PMT = 800.00; i = 3.75%; n = 120; d = 84; $c = \dfrac{2}{12} = \dfrac{1}{6}$

$$p = 1.0375^{\frac{1}{6}} - 1 = 1.0061545 - 1 = 0.61545\%$$

$$PV_{nc} = 800.00\left[\frac{1 - 1.0061545^{-120}}{0.0061545}\right] = 800.00(84.6707762) = 67\,736.62$$

$$PV_{nc}(\text{defer}) = 67\,736.62\left(1.0061545^{-84}\right) = 67\,736.62(0.5972654) = \$40\,456.74$$

5.   (a)   PMT(due) = 400.00; $i = 0.5\%$; $n = 48$; $d = 90$

$$PV_n = 400.00(1.005)\left[\frac{1 - 1.005^{-48}}{0.005}\right] = 400.00(1.005)(42.5803178) = 17\,117.29$$

$$PV_n(\text{defer}) = 17\,117.29\left(1.005^{-90}\right) = 17\,117.29(0.6383435) = \$10\,926.71$$

  (b)   Total withdrawals = 48(400.00) = \$19\,200.00

  (c)   Interest = 19\,200.00 − 10\,926.71 = \$8273.29

7.   PMT = 1200.00; $i = 2.5\%$; $n = 24$; $d = 36$

$$PV_n = 1200.00\left[\frac{1 - 1.025^{-24}}{0.025}\right] = 1200.00(17.8849858) = 21\,461.983$$

$$PV_n(\text{defer}) = 21\,461.98\left(1.025^{-36}\right) = 21\,461.98(0.4110937) = \$8822.89$$

Exercise 12.4

A.   1.   PMT = 1250.00; $i = 1.7\%$

$$PV = \frac{1250.00}{0.017} = \$73\,529.41$$

3.   PMT = 125.00; $i = 0.4375\%$

$$PV = 125.00 + \frac{125.00}{0.004375} = 125.00 + 28\,571.43 = \$28\,696.43$$

5.   PMT = 5600.00; $i = 1\%$; $c = \dfrac{12}{2} = 6$

$$p = 1.01^6 - 1 = 1.0615202 - 1 = 6.15202\%$$

$$PV = \frac{5600.00}{0.0615202} = \$91\,027.01$$

7.   PMT = 725.00(due); $i = 2.5\%$; $c = \dfrac{4}{12} = \dfrac{1}{3}$

$$p = 1.025^{\frac{1}{3}} - 1 = 1.0082648 - 1 = 0.82648\%$$

$$PV(\text{due}) = 725.00 + \frac{725.00}{0.0082648} = 725.00 + 87\,721.42 = \$88\,446.42$$

B. 1. PMT = 2500.00; i = 7.25%; d = 4; due

$$PV = \left(1.0725^{-4}\right)\left[2500.00 + \frac{2500.00}{0.0725}\right] = 0.7558068(2500.00 + 34\,482.76) = \$27\,951.82$$

3. PMT = 4.25; i = 4%; $c = \dfrac{2}{4} = \dfrac{1}{2}$

$p = 1.04^{\frac{1}{2}} - 1 = 1.0198039 - 1 = 1.98039\%$

$PV = \dfrac{4.25}{0.0198039} = \$214.60$

5. PMT = 4200.00(due); i = 0.75%; $c = \dfrac{12}{4} = 3$

$p = 1.0075^3 - 1 = 1.0226692 - 1 = 2.26692\%$

$PV = 4200.00 + \dfrac{4200.00}{0.0226692} = 4200.00 + 185\,273.41 = \$189\,473.41$

7. PV = 1400.00; i = 7%; d = 4

$PMT = \left[(1400.00)\left(1.07^4\right)\right](0.07)$

$\quad = [(1400.00)(1.3107960)(0.07)]$

$\quad = 1835.11(0.07)$

$\quad = \$128.46$

9. PV = 25 000.00; i = 1.6875%; $c = \dfrac{4}{2} = 2$

$f = 1.016875^2 - 1 = 1.0340348 - 1 = 3.40348\%$

$PMT = 25\,000.00(0.0340348) = \$850.87$

Exercise 12.5

A. 1. (a) PMT = 500.00; k = 3% = 0.03; n = 25

$PMT_{25} = PMT(1 + k)^{n-1} = 500.00(1.03)^{24}$

$\qquad\qquad\quad = 500.00(2.0327941) = \$1016.40$

(b)   Total deposited $= PMT\left\{\dfrac{(1 + k)^n - 1}{k}\right\} = 500.00\left\{\dfrac{(1.03)^{25} - 1}{0.03}\right\}$

$= 500.00(36.4592643) = \$18\,229.63$

3.   (a) $PMT = 1200.00;\ k = 1.5\% = 0.015;\ i = 2.5\%;\ n = 20$

Total deposited $= PMT\left\{\dfrac{(1 + k)^n - 1}{k}\right\} = 1200.00\left\{\dfrac{(1.015)^{20} - 1}{0.015}\right\}$

$= 1200.00(23.1236671) = \$27\,748.40$

(b) $FV = PMT\left\{\dfrac{(1 + i)^n - (1 + k)^n}{n - k}\right\} = 1200.00\left\{\dfrac{(1.025)^{20} - (1.015)^{20}}{0.025 - 0.015}\right\}$

$= 1200.00\left\{\dfrac{1.6386164 - 1.3468550}{0.01}\right\} = 1200.00\left\{\dfrac{0.2917614}{0.01}\right\}$

$= 1200.00(29.17614) = \$35\,011.37$

(c) $PMT = 1200.00;\ k = 1.5\% = 0.015;\ n = 12$

$PMT_{12} = PMT(1 + k)^{n - 1} = 1200.00(1.015)^{11}$

$= 1200.00(1.1779489) = \$1413.54$

(d)   Interest included in the accumulated value
$= 35\,011.37 - 27\,748.40 = \$7262.97$

(d)   Interest included in the total amount withdrawn
$= 58\,623.91 - 47\,588.63 = \$11\,035.28$

B.   1. $PMT = 400.00;\ k = 0.5\% = 0.005;\ i = 0.5\% = 0.005;\ n = 120$
We need to determine the present value of the withdrawals and the total amount withdrawn.

Since $i = k,\ PV = n(PMT)(1 + i)^{-1} = 120(400.00)(1.005)^{-1}$

$= 120(400.00)(0.9950249) = \$47\,761.20$

Total withdrawn $= PMT\left\{\dfrac{(1 + k)^n - 1}{k}\right\} = 400.00\left\{\dfrac{(1.005)^{120} - 1}{0.005}\right\}$

$= 400.00(163.8793468) = \$65\,551.74$

Interest included in the total amount withdrawn
$= 65\,551.74 - 47\,761.20 = \$17\,790.54$

3. $PV = 150\,000.00;\ k = -0.6\% = -0.006;\ i = 0.6\% = 0.006;\ n = 120$

$$PV = PMT\left\{\frac{1 - (1 + k)^{n}(1 + i)^{-n}}{i - k}\right\}$$

$$150\,000.00 = PMT\left\{\frac{1 - (0.994)^{120}(1.006)^{-120}}{0.006 - (-0.006)}\right\}$$

$$150\,000.00 = PMT\left\{\frac{1 - (0.4856978)(0.4878006)}{0.012}\right\}$$

$$150\,000.00 = PMT\left\{\frac{1 - 0.2369237}{0.012}\right\}$$

$$150\,000.00 = PMT(63.5896917)$$

$$PMT = \$2358.87$$

$$Total\ received = PMT\left\{\frac{(1 + k)^{n} - 1}{k}\right\} = 2358.87\left\{\frac{(0.994)^{120} - 1}{-0.006}\right\}$$

$$= 2358.87(85.7170349) = \$202\,195.34$$

5. $FV = 500\,000.00;\ k = 2\% = 0.02;\ i = 3.5\% = 0.035;\ n = 70$

$$FV = PMT\left\{\frac{(1 + i)^{n} - (1 + k)^{n}}{i - k}\right\}$$

$$500\,000.00 = PMT\left\{\frac{(1.035)^{70} - (1.02)^{70}}{0.035 - 0.02}\right\}$$

$$500\,000.00 = PMT\left\{\frac{11.1128253 - 3.9995582}{0.015}\right\}$$

$$500\,000.00 = PMT\left\{\frac{7.1132671}{0.015}\right\}$$

$$500\,000.00 = PMT(474.2178)$$

$$PMT = \$1054.37$$

$$Total\ amount = PMT\left\{\frac{(1 + k)^{n} - 1}{k}\right\} = 1054.37\left\{\frac{(1.02)^{70} - 1}{0.02}\right\}$$

$$= 1054.37(149.9779111) = \$158\,132.21$$

7. PMT = 400.00; $k = 1.25\% = 0.0125$; $i = 2.5\% = 0.025$; $n = 70$
Value of fund after the last semi-annual payment:

$$FV = PMT\left\{\frac{(1 + i)^n - (1 + k)}{i - k}\right\} = 400.00\left\{\frac{(1.025)^{70} - (1.0125)^{70}}{0.025 - 0.0125}\right\}$$

$$= 400.00\left\{\frac{5.6321029 - 2.3859000}{0.0125}\right\} = 400.00\left\{\frac{3.2462029}{0.0125}\right\}$$

$$= 400.00(259.696232) = \$103\,878.49$$

Value of fund five years after the last deposit:
PV = 103 878.49; $i = 1.5\% = 0.015$; $n = 20$

$$FV = PV(1 + i)^n = 103\,878.49(1.015)^{20}$$

$$= 103\,878.49(1.3468550) = \$139\,909.26$$

Size of first withdrawal
PV = 139 909.26; $k = 2\% = 0.02$; $i = 1.5\% = 0.015$; $n = 80$

$$PV = PMT\left\{\frac{1 - (1 + k)^n(1 + i)^{-n}}{i - k}\right\}$$

$$139\,909.26 = PMT\left\{\frac{1 - (1.02)^{80}(1.015)^{-80}}{0.015 - 0.02}\right\}$$

$$139\,909.26 = PMT\left\{\frac{1 - (4.8754392)(0.3038901)}{-0.005}\right\}$$

$$139\,909.26 = PMT\left\{\frac{1 - 1.4815977}{-0.005}\right\}$$

$$139\,909.26 = PMT(96.319540)$$

$$PMT = \$1452.55$$

$$Total\ withdrawn = PMT\left\{\frac{(1 + k)^n - 1}{k}\right\} = 1452.55\left\{\frac{(1.02)^{80} - 1}{0.02}\right\}$$

$$= 1452.55(193.7719578) = \$281\,463.46$$

Review Exercise

1. PMT = 540.00; $i = 4.5\%$; $n = 15$

(a) $FV_n = 540.00\left[\frac{1.045^{15} - 1}{0.045}\right] = 540.00(20.7840543) = \$11\,223.39$

   $PV_n = 540.00\left[\frac{1 - 1.045^{-15}}{0.045}\right] = 540.00(10.7395457) = \$5799.35$

(b) $FV_n(due) = 540.00\left[\frac{1.045^{15} - 1}{0.045}\right](1.045) = 11\,223.39(1.045) = 11\,728.44$

   $PV_n(due) = 540.00\left[\frac{1 - 1.045^{-15}}{0.045}\right](1.045) = 5799.35(1.045) = 6060.32$

3. PMT $= 375.00$; $i = 0.3125\%$; $n = 32$; $c = \dfrac{12}{4} = 3$

$p = 1.003125^3 - 1 = 1.0094043 - 1 = 0.94043\%$

   (a) $FV_{nc} = 375.00\left[\dfrac{1.0094043^{32} - 1}{0.0094043}\right] = 375.00(37.1347524) = \$13\,925.53$

   (b) $FV_{nc}(\text{due}) = 375.00(1.0094043)\left[\dfrac{1.0094043^{32} - 1}{0.0094043}\right]$

                    $= 13\,925.53(1.0094043) = \$14\,056.49$

5. PMT $= 368.00$; $i = 0.875\%$; $n = 180$

$PV_n = 368.00\left[\dfrac{1 - 1.00875^{-180}}{0.00875}\right] = 368.00(90.4650781) = \$33\,291.15$

7. PMT $= 3500.00$; $i = 2\%$; $c = \dfrac{4}{2} = 2$

$p = 1.02^2 - 1 = 1.0404 - 1 = 4.04\%$

   (a) $n = 30$

        $PV_{nc} = 3500.00\left[\dfrac{1 - 1.0404^{-30}}{0.0404}\right]$

              $= 3500.00(17.2083597) = \$60\,229.26$

   (b) $n = 20$; due

        $PV_{nc}(\text{due}) = 3500.00(1.0404)\left[\dfrac{1 - 1.0404^{-20}}{0.0404}\right]$

                    $= 3500.00(1.0404)(13.5423165)$

                    $= \$49\,312.99$

   (c) $n = 16$; $d = 8$

        $PV_{nc} = 3500.00\left[\dfrac{1 - 1.0404^{-16}}{0.0404}\right] = 3500.00(11.6179875) = 40\,662.96$

        $PV_{nc}(\text{defer}) = 40\,662.96\left(1.0404^{-8}\right) = 40\,662.96(0.7284458) = \$29\,620.76$

(d) $n = 18$; $d = 6$

$$PV_{nc}(\text{due}) = 3500.00(1.0404)\left[\frac{1 - 1.0404^{-18}}{0.0404}\right]$$

$$= 3500.00(1.0404)(12.6182389) = 45\,948.06$$

$$PV_{nc}(\text{defer}) = 45\,948.06\left(1.0404^{-6}\right)$$

$$= 45\,948.06(0.7884932) = \$36\,229.73$$

(e) $$PV = \frac{3500.00}{0.0404} = \$86\,633.66$$

(f) $$PV(\text{due}) = 3500.00 + \frac{3500.00}{0.0404} = 3500.00 + 86\,633.66 = \$90\,133.66$$

9. $PMT = 25.00$; $i = 1\%$; $n = 360$; $c = \dfrac{4}{12} = \dfrac{1}{3}$

$$p = 1.01^{\frac{1}{3}} - 1 = 1.0033223 - 1 = 0.33223\%$$

$$FV_{nc}(\text{due}) = 25.00(1.0033223)\left[\frac{1.0033223^{360} - 1}{0.0033223}\right]$$

$$= 25.00(1.0033223)(692.4113805)$$

$$= \$17\,367.86$$

11. $PMT = 4800.00$; $i = 1.5\%$; $n = 80$; $d = 40$; due

$$PV_n(\text{defer}) = 4800.00\left(1.015^{-40}\right)\left[\frac{1 - 1.015^{-80}}{0.015}\right](1.015)$$

$$= 4800.00\left(1.015^{-39}\right)\left[\frac{1 - 1.015^{-80}}{0.015}\right]$$

$$= 4800.00(0.5595313)(46.4073235)$$

$$= \$124\,638.48$$

13. $PMT = 800.00$; $i = 2.875\%$; $n = 240$; $d = 180$; $c = \dfrac{2}{12} = \dfrac{1}{6}$

$$p = 1.02875^{\frac{1}{6}} - 1 = 1.0047353 - 1 = 0.47353\%$$

$$PV_{nc} = 800.00\left[\frac{1 - 1.0047353^{-240}}{0.0047353}\right] = 800.00(143.2197801) = 114\,575.82$$

$$PV_{nc}(\text{defer}) = 114\,575.82\left(1.0047353^{-180}\right)$$

$$= 114\,575.82(0.4272687) = \$48\,954.66$$

15. PMT = 1750.00; $i = 0.55\%$

$$PV = \frac{1750.00}{0.0055} = \$318\,181.82$$

17. PMT = 7.50; $i = 3.5\%$; $c = \frac{4}{2} = 2$

$p = 1.035^2 - 1 = 1.071225 - 1 = 7.1225\%$

$$PV = \frac{7.50}{0.071225} = \$105.30$$

19. (a) PMT = 82.00; $i = 1.375\%$; $n = 42$

$$PV_n(\text{due}) = 82.00(1.01375)\left[\frac{1 - 1.01375^{-42}}{0.01375}\right]$$

$$= 82.00(1.01375)(31.7445433) = \$2638.84$$

(b) Amount paid = 42(82.00) = \$3444.00

(c) Interest = 3444.00 − 2638.84 = \$805.16

21. PMT = 1200.00; $k = 1.5\% = 0.015$; $i = 1.75\% = 0.0175$; $n = 80$

$$\text{Total deposited} = \text{PMT}\left\{\frac{(1 + k)^n - 1}{k}\right\} = 1200.00\left\{\frac{(1.015)^{80} - 1}{0.015}\right\}$$

$$= 1200.00(152.710825) = \$183\,253.02$$

$$FV = \text{PMT}\left\{\frac{(1 + i)^n - (1 + k)^n}{i - k}\right\} = 1200.00\left\{\frac{(1.0175)^{80} - (1.015)^{80}}{0.0175 - 0.015}\right\}$$

$$= 1200.00\left\{\frac{4.0063919 - 3.2906628}{0.0025}\right\} = 1200.00\left\{\frac{0.7157291}{0.0025}\right\}$$

$$= 1200.00(286.29164) = \$343\,549.97$$

Interest = 343 549.97 − 183 253.02 = \$160 296.95

Self-Test

1. PMT = 1080.00; $i = 2.375\%$; $n = 44$

$$FV_n(\text{due}) = 1080.00(1.02375)\left[\frac{1.02375^{44} - 1}{0.02375}\right]$$

$$= 1080.00(1.02375)(76.1631135)$$

$$= \$84\,209.75$$

3. PMT = 960.00; i = 0.5%; n = 84

$$PV_n(\text{due}) = 960.00(1.005)\left[\frac{1 - 1.005^{-84}}{0.005}\right]$$

$$= 960.00(1.005)(68.4530424)$$

$$= 65\,714.92(1.005)$$

$$= \$66\,043.50$$

5. PMT = 0.75; i = 1.125%; $c = \dfrac{12}{4} = 3$

$p = 1.01125^3 - 1 = 1.0341311 - 1 = 0.0341311 = 3.41311\%$

$PV = \dfrac{0.75}{0.0341311} = \$21.97$

7. PMT = 480; i = 5.5%

$$PV(\text{due}) = 480.00 + \frac{480.00}{0.055} = 480.00 + 8727.27 = \$9207.27$$

9. PMT = 3000.00; i = 7.0%; d = 4

$$PV = \frac{3000.00}{0.07}\left(1.07^{-4}\right) = 42\,857.14(0.7628952) = \$32\,695.51$$

11. PV = 250 000.00; k = 1.75% = 0.0175; i = 4% = 0.04; n = 40

$$PV = PMT\left\{\frac{1 - (1 + k)^n(1 + i)^{-n}}{i - k}\right\}$$

$$250\,000.00 = PMT\left\{\frac{1 - (1.0175)^{40}(1.04)^{-40}}{0.04 - 0.175}\right\}$$

$$250\,000.00 = PMT\left\{\frac{1 - (2.0015973)(0.2082890)}{0.0225}\right\}$$

$$250\,000.00 = PMT\left\{\frac{1 - 0.4169108}{0.0225}\right\}$$

$$250\,000.00 = PMT(25.915076)$$

$$PMT = \$9646.89$$

$$\text{Total withdrawn} = PMT\left\{\frac{(1 + k)^n - 1}{k}\right\} = 9646.89\left\{\frac{(1.0175)^{40} - 1}{0.0175}\right\}$$

$$= 9646.89(57.234134) = \$552\,131.39$$

Interest included = 552 131.39 − 250 000.00 = \$302 131.39

13 Annuities—Finding R, n, or i

Exercise 13.1

A.   1. $FV_n = 15\,000.00$; $i = 2.75\%$; $n = 15$

$$15\,000.00 = PMT\left[\frac{1.0275^{15} - 1}{0.0275}\right]$$

$15\,000.00 = 18.26178052PMT$

     $PMT = \$821.39$

3. $PV_n = 12\,000.00$; $i = 4.5\%$; $n = 15$

$$12\,000.00 = PMT\left[\frac{1 - 1.045^{-15}}{0.045}\right]$$

$12\,000.00 = 10.739547PMT$

     $PMT = \$1117.37$

5. $FV_n = 8000.00$; $i = 1.7\%$; $n = 24$

$$8000.00 = PMT\left[\frac{1.017^{24} - 1}{0.017}\right]$$

$8000.00 = 29.3328913PMT$

     $PMT = \$272.73$

7. $FV_{nc} = 45\,000.00$; $i = 0.75\%$; $n = 40$; $c = \dfrac{12}{4} = 3$

   $p = 1.0075^3 - 1 = 1.0226692 - 1 = 2.26692\%$

$$45\,000.00 = PMT\left[\frac{1.0226692^{40} - 1}{0.0226692}\right]$$

$45\,000.00 = 64.0234227PMT$

     $PMT = \$702.87$

9. $PV_{nc} = 20\,000.00$; $i = 3.5\%$; $n = 96$; $c = \dfrac{2}{12} = \dfrac{1}{6}$

   $p = 1.035^{\frac{1}{6}} - 1 = 1.0057500 - 1 = 0.575\%$

$$20\,000.00 = PMT\left[\frac{1 - 1.00575^{-96}}{0.00575}\right]$$

$20\,000.00 = 73.6159851PMT$

     $PMT = \$271.68$

B. 1. $FV_n(\text{due}) = 20\,000.00$; $i = 1.5\%$; $n = 60$

$$20\,000.00 = PMT(1.015)\left[\frac{1.015^{60} - 1}{0.015}\right]$$

$$20\,000.00 = PMT(1.015)(96.2146517)$$

$$PMT = \$204.80$$

3. $PV_n(\text{due}) = 18\,500.00$; $i = 1.5\%$; $n = 24$

$$18\,500.00 = PMT(1.015)\left[\frac{1 - 1.015^{-24}}{0.015}\right]$$

$$18\,500.00 = PMT(1.015)(20.0304054)$$

$$PMT = \$909.95$$

5. $FV_{nc}(\text{due}) = 16\,500.00$; $i = 1\%$; $n = 10$; $c = \dfrac{4}{1} = 4$

$$p = 1.01^4 - 1 = 1.040604 - 1 = 4.0604\%$$

$$16\,500.00 = PMT(1.040604)\left[\frac{1.040604^{10} - 1}{0.040604}\right]$$

$$16\,500.00 = PMT(1.040604)(12.039789)$$

$$PMT = \$1316.98$$

7. $PV_{nc}(\text{due}) = 10\,000.00$; $i = 0.5\%$; $n = 12$; $c = \dfrac{12}{4} = 3$

$$p = 1.005^3 - 1 = 1.0150751 - 1 = 1.50751\%$$

$$10\,000.00 = PMT(1.0150751)\left[\frac{1 - 1.0150751^{-12}}{0.0150751}\right]$$

$$10\,000.00 = PMT(1.0150751)(10.9024042)$$

$$PMT = \$903.61$$

C. 1. $PV_n(\text{defer}) = 7200.00$; $i = 2.25\%$; $n = 20$; $d = 12$

$$7200.00 = PV_n\left(1.0225^{-12}\right)$$

$$7200.00 = 0.7656675 PV_n$$

$$PV_n = 9403.560$$

$$9403.560 = PMT(\text{due})(1.0225)\left[\frac{1 - 1.0225^{-20}}{0.0225}\right]$$

$$9403.560 = PMT(\text{due})(1.0225)(15.9637124)$$

$$PMT(\text{due}) = \$576.10$$

3.　$PV_{nc}(\text{defer}) = 9000.00;\ i = 1.25\%;\ n = 84;\ d = 36;\ c = \dfrac{4}{12} = \dfrac{1}{3}$

$$p = 1.0125^{\frac{1}{3}} - 1 = 1.0041494 - 1 = 0.41494\%$$

$$9000.00 = PV_{nc}\left(1.0041494^{-36}\right)$$

$$9000.00 = 0.8615094\,PV_{nc}$$

$$PV_{nc} = 10\,446.78096$$

$$10\,446.78096 = PMT\left[\frac{1 - 1.0041494^{-84}}{0.0041494}\right]$$

$$10\,446.78096 = 70.8005948\,PMT$$

$$PMT = \$147.55$$

D.　1.　$FV_n = 20\,000.00;\ i = 1.5\%;\ n = 60$

$$20\,000.00 = PMT\left[\frac{1.015^{60} - 1}{0.015}\right]$$

$$20\,000.00 = 96.2146517\,PMT$$

$$PMT = \$207.87$$

3.　$PV_n = 7500.00;\ i = 4.8\%;\ n = 20$

$$7500.00 = PMT\left[\frac{1 - 1.048^{-20}}{0.048}\right]$$

$$7500.00 = 12.6762836\,PMT$$

$$PMT = \$591.66$$

5.　$PV_{nc} = 32\,000.00;\ i = 4.75\%;\ n = 180;\ c = \dfrac{2}{12} = \dfrac{1}{6}$

$$p = 1.0475^{\frac{1}{6}} - 1 = 1.0077644 - 1 = 0.77644\%$$

$$32\,000.00 = PMT\left[\frac{1 - 1.0077644^{-180}}{0.0077644}\right]$$

$$32\,000.00 = 96.7841192\,PMT$$

$$PMT = \$330.63$$

7. $FV_{nc} = 20\,000.00;\ i = 3.5\%;\ n = 80;\ c = \dfrac{2}{4} = \dfrac{1}{2}$

$p = 1.035^{\frac{1}{2}} - 1 = 1.01734950 - 1 = 1.73495\%$

$20\,000.00 = PMT\left[\dfrac{1.0173495^{80} - 1}{0.0173495}\right]$

$20\,000.00 = 170.5674809 PMT$

$\qquad PMT = \$117.26$

9. $PV_n(\text{due}) = 12\,500.00;\ i = 0.625\%;\ n = 48$

$12\,500.00 = PMT(1.00625)\left[\dfrac{1 - 1.00625^{-48}}{0.00625}\right]$

$12\,500.00 = PMT(1.00625)(41.3583711)$

$\qquad PMT = \$300.36$

11. $FV_{nc}(\text{due}) = 10\,000.00;\ i = 1.75\%;\ n = 16;\ c = \dfrac{4}{2} = 2$

$p = 1.0175^{2} - 1 = 1.0353063 - 1 = 3.53063\%$

$10\,000.00 = PMT(1.0353063)\left[\dfrac{1.0353063^{16} - 1}{0.0353063}\right]$

$10\,000.00 = PMT(1.0353063)(21.0221643)$

$\qquad PMT = \$459.47$

13. $PV_n(\text{defer}) = 20\,000.00;\ i = 2.375\%;\ n = 24;\ d = 40$

$20\,000.00 = PV_n\left(1.02375^{-40}\right)$

$20\,000.00 = 0.391060 PV_n$

$\qquad PV_n = 51\,143.05$

$51\,143.05 = PMT\left[\dfrac{1 - 1.02375^{-24}}{0.02375}\right]$

$51\,143.05 = 18.1344685 PMT$

$\qquad PMT = \$2820.21$

15. $PV_{nc}(\text{defer}) = 18\,000.00;\ i = 1.75\%;\ n = 16;\ d = 4;\ c = \dfrac{4}{2} = 2$

$p = 1.0175^2 - 1 = 1.03530625 - 1 = 3.530625\%$

$18\,000.00 = PV_{nc}(\text{due})\left(1.03530625^{-4}\right)$

$18\,000.00 = 0.8704116\,PV_{nc}(\text{due})$

$PV_{nc}(\text{due}) = 20\,679.87$

$20\,679.87 = PMT(\text{due})(1.03530625)\left[\dfrac{1 - 1.03530625^{-16}}{0.03530625}\right]$

$20\,679.87 = PMT(\text{due})(1.03530625)(12.0663489)$

$PMT(\text{due}) = \$1655.40$

17. $PV_n = 0.90(10\,600.00) = 9540.00;\ i = 0.6\%;\ n = 48$

$9540.00 = PMT\left[\dfrac{1 - 1.006^{-48}}{0.006}\right]$

$9540.00 = 41.5988193\,PMT$

$PMT = \$229.33$

19. Amount after 15 years:
$PMT = 1200.00;\ i = 3.75\%;\ n = 30$

$FV_n = 1200.00\left[\dfrac{1.0375^{30} - 1}{0.0375}\right]$

$= 1200.00(53.7992372)$

$= \$64\,559.08$

RRIF withdrawals:
$PV_n = 64\,559.08;\ i = 3.75\%;\ n = 40$

$64\,559.08 = PMT\left[\dfrac{1 - 1.0375^{-40}}{0.0375}\right]$

$64\,559.08 = 20.5509900\,PMT$

$PMT = \$3141.41$

21. Amount after 18 years:
$$PV = 2000.00; \quad i = 1.25\%; \quad n = 72$$
$$\begin{aligned} FV &= 2000.00(1.0125)^{72} \\ &= 2000.00(2.4459203) \\ &= \$4891.84 \end{aligned}$$
Withdrawals:
$$PV_n = 4891.84; \quad i = 0.5\%; \quad n = 48$$
$$4891.84 = PMT\left[\frac{1 - 1.005^{-48}}{0.005}\right]$$
$$4891.84 = 42.5803178PMT$$
$$PMT = \$114.89$$

23. Amount after 20 years:
$$PMT = 1500.00; \quad i = 3.125\%; \quad n = 40$$
$$\begin{aligned} FV_n(due) &= 1500.00(1.03125)\left[\frac{1.03125^{40} - 1}{0.03125}\right] \\ &= 1500.00(1.03125)(77.5742333) \\ &= \$119\,997.64 \end{aligned}$$
Annuity due payment:
$$PV_n(due) = 119\,997.64; \quad i = 0.55\%; \quad n = 180$$
$$119\,997.64 = PMT(1.0055)\left[\frac{1 - 1.0055^{-180}}{0.0055}\right]$$
$$119\,997.64 = PMT(1.0055)(114.0752759)$$
$$PMT = \$1046.16$$

25. $$PV_n(defer) = 50\,000.00; \quad i = 1.75\%; \quad n = 28; \quad d = 12$$
$$50\,000.00 = PV_n\left(1.0175^{-12}\right)$$
$$50\,000.00 = 0.8120579PV_n$$
$$PV_n = 61\,571.96$$
$$61\,571.96 = PMT(due)(1.0175)\left[\frac{1 - 1.0175^{-28}}{0.0175}\right]$$
$$61\,571.96 = PMT(due)(1.0175)(21.9869547)$$
$$PMT(due) = \$2752.22$$

Exercise **13.2**

1. $FV_n = 20\,000.00$; $PMT = 800.00$; $i = 7.5\%$

$$20\,000.00 = 800.00\left[\frac{1.075^n - 1}{0.075}\right]$$

$$25.00 = \frac{1.075^n - 1}{0.075}$$

$$1.875 = 1.075^n - 1$$

$$1.075^n = 2.875$$

$$n\ln 1.075 = \ln 2.875$$

$$0.0723207n = 1.0560527$$

$$n = 14.6023573 \text{ years}$$

$$n = 14 \text{ years } 8 \text{ months}$$

3. $PV_n = 14\,500.00$; $PMT = 190.00$; $i = 0.4375\%$

$$14\,500.00 = 190.00\left[\frac{1 - 1.004375^{-n}}{0.004375}\right]$$

$$76.3157895 = \frac{1 - 1.004375^{-n}}{0.004375}$$

$$0.3338816 = 1 - 1.004375^{-n}$$

$$1.004375^{-n} = 0.6661184$$

$$-n\ln 1.004375 = \ln 0.6661184$$

$$-0.00436546n = -0.40628785$$

$$n = 93.0687373 \text{ months}$$

$$n = 7 \text{ years } 10 \text{ months}$$

5. $FV_n = 3600.00$; $PMT = 175.00$; $i = 3.7\%$

$$3600.00 = 175.00\left[\frac{1.037^n - 1}{0.037}\right]$$

$$20.5714286 = \frac{1.037^n - 1}{0.037}$$

$$0.76114286 = 1.037^n - 1$$

$$1.037^n = 1.76114286$$

$$n\ln 1.037 = \ln 1.76114286$$

$$0.0363319n = 0.5659630$$

$$n = 15.5775778 \text{ half years}$$

$$n = 7 \text{ years } 10 \text{ months}$$

7. $PV_{nc} = 21\,400.00$; $PMT = 1660.00$; $i = 0.375\%$; $c = \dfrac{12}{2} = 6$

$p = 1.00375^6 - 1 = 1.0227120 - 1 = 2.27120\%$

$$21\,400.00 = 1660.00\left[\frac{1 - 1.0227120^{-n}}{0.0227120}\right]$$

$$12.8915663 = \frac{1 - 1.0227120^{-n}}{0.0227120}$$

$$0.2927933 = 1 - 1.0227120^{-n}$$

$$1.0227120^{-n} = 0.7072067$$

$-n \ln 1.0227120 = \ln 0.7072067$

$-0.0224579n = -0.3464323$

$\qquad\qquad n = 15.4258546$ half-years

$\qquad\qquad n = 7$ years 9 months

9. $FV_{nc} = 7200.00$; $PMT = 90.00$; $i = 1.875\%$; $c = \dfrac{2}{12} = \dfrac{1}{6}$

$f = 1.01875^{\frac{1}{6}} - 1 = 1.0031009 - 1 = 0.31009\%$

$$7200.00 = 90.00\left[\frac{1.0031009^n - 1}{0.0031009}\right]$$

$$80.00 = \frac{1.0031009^n - 1}{0.0031009}$$

$$1.0031009^n = 1.248072$$

$n \ln 1.0031009 = \ln 1.248072$

$0.0030961n = 0.221600$

$\qquad\qquad n = 71.5739156$ months

$\qquad\qquad n = 6$ years

B.   1. $FV_n(\text{due}) = 5300.00$; $PMT = 35.00$; $i = 0.5\%$

$$5300.00 = 35.00(1.005)\left[\frac{1.005^n - 1}{0.005}\right]$$

$$150.6751955 = \frac{1.005^n - 1}{0.005}$$

$$1.005^n = 1.7533760$$

$n \ln 1.005 = \ln 1.7533760$

$0.004987542n = 0.5615431$

$\qquad\qquad n = 112.5891471$ months

The term is 113 months (9 years 5 months).

3.　$PV_n(\text{due}) = 6450.00$;　$PMT = 1120.00$;　$i = 10\%$

$$6450.00 = 1120.00(1.10)\left[\frac{1 - 1.10^{-n}}{0.10}\right]$$

$$5.2353896 = \frac{1 - 1.10^{-n}}{0.10}$$

$$1.10^{-n} = 0.476461$$

$$-n\ln 1.10 = \ln 0.476461$$

$$-0.0953102n = -0.7413694$$

$$n = 7.7784961 \text{ years}$$

The term is 8 years.

5.　$FV_n(\text{due}) = 32\,000.00$;　$PMT = 450.00$;　$i = 0.625\%$;　$c = \dfrac{12}{2} = 6$

$$p = 1.00625^6 - 1 = 1.0380908 - 1 = 3.80908\%$$

$$32\,000.00 = 450.00(1.0380908)\left[\frac{1.0380908^n - 1}{0.0380908}\right]$$

$$68.501822 = \frac{1.0380908^n - 1}{0.0380908}$$

$$1.0380908^n = 3.6092892$$

$$n\ln 1.0380908 = \ln 3.6092892$$

$$0.0373833n = 1.2835109$$

$$n = 34.3338041 \text{ half - years}$$

The term is 35 semi-annual periods (17.5 years).

7.　$PV_{nc}(\text{due}) = 12\,500.00$;　$PMT = 860.00$;　$i = 0.75\%$;　$c = \dfrac{12}{4} = 3$

$$p = 1.0075^3 - 1 = 1.0226692 - 1 = 2.26692\%$$

$$12\,500.00 = 860.00(1.0226692)\left[\frac{1 - 1.0226692^{-n}}{0.0226692}\right]$$

$$14.2126933 = \frac{1 - 1.0226692^{-n}}{0.0226692}$$

$$1.0226692^{-n} = 0.6778096$$

$$-n\ln 1.0226692 = \ln 0.6778096$$

$$-0.0224161n = -0.3888888$$

$$n = 17.3486378 \text{ quarters}$$

The term is 18 quarters (4.5 years).

C. 1. $\text{PV}_n(\text{defer}) = 21\,000.00$; $\text{PMT(due)} = 3485.00$; $i = 2.5\%$; $d = 32$

$$21\,000.00 = \text{PV}_n\left(1.025^{-32}\right)$$
$$21\,000.00 = 0.4537706\text{PV}_n$$
$$\text{PV}_n = 46\,278.89$$
$$46\,278.89 = 3485.00(1.025)\left[\frac{1 - 1.025^{-n}}{0.025}\right]$$
$$0.3238891 = 1 - 1.025^{-n}$$
$$1.025^{-n} = 0.6761109$$
$$-n\ln 1.025 = \ln 0.6761109$$
$$-0.02469261n = -0.3913982$$
$$n = 15.8508234 \text{ quarters}$$

The term is 16 quarters, i.e. 4 years.

3. $\text{PV}_{nc}(\text{defer}) = 22\,750.00$; $\text{PMT} = 385.00$; $i = 2.5\%$; $d = 12$; $c = \frac{2}{12} = \frac{1}{6}$

$$p = 1.025^{\frac{1}{6}} - 1 = 1.0041239 - 1 = 0.41239\%$$
$$22\,750.00 = \text{PV}_{nc}\left(1.0041239^{-12}\right)$$
$$22\,750.00 = 0.9518146\text{PV}_{nc}$$
$$\text{PV}_{nc} = 23\,901.71$$
$$23\,901.71 = 385.00\left[\frac{1 - 1.0041239^{-n}}{0.0041239}\right]$$
$$0.2560215 = 1 - 1.0041239^{-n}$$
$$1.0041239^{-n} = 0.7439785$$
$$-n\ln 1.0041239 = \ln 0.7439785$$
$$-0.0041154n = -0.2957431$$
$$n = 71.862 \text{ months}$$

The term is 72 months.

D. 1. $\text{FV}_n = 4500.00$; $\text{PMT} = 50.00$; $i = 0.50\%$

$$4500.00 = 50.00\left[\frac{1.005^n - 1}{0.005}\right]$$
$$90.00 = \frac{1.005^n - 1}{0.005}$$
$$1.005^n = 1.45$$
$$n\ln 1.005 = \ln 1.45$$
$$0.004987542n = 0.371563556$$
$$n = 74.4983312 \text{ months}$$

It will take 6 years 3 months.

3. $FV_{nc} = 15\,000.00;\ PMT = 90.00;\ i = 1\%;\ c = \dfrac{4}{12} = \dfrac{1}{3}$

$p = 1.01^{\frac{1}{3}} - 1 = 1.0033223 - 1 = 0.33223\%$

$$15\,000.00 = 90.00\left[\frac{1.0033223^n - 1}{0.0033223}\right]$$

$$166.666667 = \frac{1.0033223^n - 1}{0.0033223}$$

$1.0033223^n = 1.5537167$

$n \ln 1.0033223 = \ln 1.5537167$

$0.0033168n = 0.4406499$

$\qquad\qquad n = 132.854$ months

It will take 133 months.

5. $PV_n(\text{due}) = 25\,000.00;\ PMT = 1445.00;\ i = 2\%$

$$25\,000.00 = 1445.00(1.02)\left[\frac{1 - 1.02^{-n}}{0.02}\right]$$

$$16.961802 = \frac{1 - 1.02^{-n}}{0.02}$$

$1.02^{-n} = 0.6607640$

$-n \ln 1.02 = \ln 0.6607640$

$-0.0198026n = -0.4143585$

$\qquad\qquad n = 20.9244493$ quarters

The term of the lease is 21 quarters, i.e., 5 years, 3 months.

7. $PV_{nc}(\text{due}) = 85\,000.00;\ i = 6.125\%;\ PMT = 3000.00;\ c = \dfrac{1}{4}$

$p = 1.06125^{\frac{1}{4}} - 1 = 1.0149729 - 1 = 1.49729\%$

$$85\,000.00 = 3000.00(1.0149729)\left[\frac{1 - 1.0149729^{-n}}{0.0149729}\right]$$

$$27.915359 = \frac{1 - 1.0149729^{-n}}{0.0149729}$$

$1.0149729^{-n} = 1 - 0.4179739$

$-n \ln 1.0149729 = \ln 0.5820261$

$-0.0148619n = -0.541240$

$\qquad\qquad n = 36.418$ quarters

Withdrawals can be made for 37 quarters, i.e., 9 years, 3 months.

9. Amount after 5 years:
    PV $= 4000.00$; $i = 1\%$; $n = 20$

    FV $= 4000.00(1.01)^{20}$
    $= 4000.00(1.220190)$
    $= \$4880.76$

    Withdrawals:
    $PV_n = 4880.76$; PMT $= 500.00$; $i = 1\%$

    $$4880.76 = 500.00\left[\frac{1 - 1.01^{-n}}{0.01}\right]$$

    $$9.76152 = \frac{1 - 1.01^{-n}}{0.01}$$

    $$1.01^{-n} = 0.9023848$$
    $$-n\ln 1.01 = \ln 0.9023848$$
    $$-0.00995033n = -0.1027142$$
    $$n = 10.3227 \text{ quarters}$$

    Ten withdrawals of $500.00 and a final withdrawal of less than $500 can be made.

11. Amount after 10 years:
    PMT $= 75.00$; $i = 0.5\%$; $n = 120$

    $$FV_n(\text{due}) = 75.00(1.005)\left[\frac{1.005^{120} - 1}{0.005}\right]$$
    $$= 75.00(1.005)(163.879347)$$
    $$= \$12\,352.41$$

    Withdrawals:
    $PV_n(\text{due}) = 12\,352.41$; PMT $= 260.00$; $i = 0.5\%$

    $$12\,352.41 = 260.00(1.005)\left[\frac{1 - 1.005^{-n}}{0.005}\right]$$

    $$47.272905 = \frac{1 - 1.005^{-n}}{0.005}$$

    $$1.005^{-n} = 0.7636355$$
    $$-n\ln 1.005 = \ln 0.7636355$$
    $$-0.00498754n = -0.2696647$$
    $$n = 54.068 \text{ months}$$

    Withdrawals can be made for 4 years 7 months.

13. $PV_n(\text{defer}) = 16\,000.00$; $PMT = 1000.00$; $i = 0.6\%$; $d = 72$

$$16\,000.00 = PV_n\left(1.006^{-72}\right)$$
$$16\,000.00 = 0.6500479 PV_n$$
$$PV_n = 24\,613.57$$
$$24\,613.57 = 1000.00\left[\frac{1 - 1.006^{-n}}{0.006}\right]$$
$$0.1476814 = 1 - 1.006^{-n}$$
$$1.006^{-n} = 0.8523186$$
$$-n\ln 1.006 = \ln 0.8523186$$
$$-0.00598207n = -0.1597949$$
$$n = 26.712 \text{ months}$$

15. $PV_n(\text{defer}) = 6500.00$; $PMT = 300.00$; $i = 0.7\%$; $d = 48$

$$6500.00 = PV_n\left(1.007^{-48}\right)$$
$$6500.00 = 0.7154601 PV_n$$
$$PV_n = 9085.063$$
$$9085.063 = 300.00\left[\frac{1 - 1.007^{-n}}{0.007}\right]$$
$$0.2119848 = 1 - 1.007^{-n}$$
$$1.007^{-n} = 0.7880152$$
$$-n\ln 1.007 = \ln 0.7880152$$
$$-0.006975614n = -0.2382379$$
$$n = 34.152966 \text{ months}$$

17. $PV_{nc} = 18\,600.00$; $PMT = 260.00$; $i = 4\%$; $c = \dfrac{2}{12} = \dfrac{1}{6}$

$$p = 1.04^{\frac{1}{6}} - 1 = 1.0065582 - 1 = 0.65582\%$$
$$18\,600.00 = 260.00\left[\frac{1 - 1.0065582^{-n}}{0.0065582}\right]$$
$$71.5384615 = \frac{1 - 1.0065582^{-n}}{0.0065582}$$
$$1.0065582^{-n} = 0.5308365$$
$$-n\ln 1.0065582 = \ln 0.5308365$$
$$-0.00653679n = -0.63330122$$
$$n = 96.882601 \text{ months}$$

19. $PV_{nc}(\text{defer}) = 45\,000.00$; $PMT = 15\,000.00$; $i = 5.5\%$; $d = 4$; $c = 2$

$$p = 1.055^2 - 1 = 1.113025 - 1 = 11.3025\%$$

$$45\,000.00 = PV_{nc}\left(1.113025^{-4}\right)$$

$$45\,000.00 = 0.6515989PV_{nc}$$

$$PV_{nc} = 69\,060.89$$

$$69\,060.89 = 15\,000\left[\frac{1 - 1.113025^{-n}}{0.113025}\right]$$

$$0.5203738 = 1 - 1.113025^{-n}$$

$$1.113025^{-n} = 0.4796262$$

$$-n\ln 1.113025 = \ln 0.4796262$$

$$-0.1070815n = -0.7347482$$

$$n = 6.8615793 \text{ years}$$

Exercise 13.3

A. 1. $FV_n = 9000.00$; $PMT = 230.47$; $n = 32$; $m = 4$

0	9000.00 or (−9000.00)	230.47	32		
PV	FV	PMT	N	CPT	I/Y

$i = 1.25003\%$
The nominal annual rate is $1.25003\% \times 4 = 5\%$.

3. $PV_n = 7400.00$; $PMT = 119.06$; $n = 84$; $m = 12$

0	7400.00	119.06	84		
FV	PV	PMT	N	CPT	I/Y

$i = 0.75\%$
The nominal annual rate is $0.75\% \times 12 = 9.0\%$.

5. $FV_n(\text{due}) = 70\,000.00$; $PMT = 1014.73$; $i = 25$; $m = 1$

0	70 000.00 or (−70 000.00)	1014.73	25					
PV	FV	PMT	N	DUE	(or 2nd	CPT)		I/Y

$i = 7.1250080\%$
The nominal annual rate is 7.125%.

7. $PV_n(\text{due}) = 28\,700.00$; $PMT = 2015.00$; $n = 30$; $m = 2$

0	28 700.00	2015.00	30					
FV	PV	PMT	N	DUE	(or	2nd	CPT)	I/Y

$i = 6.2492018\%$
The nominal annual rate is 12.5%.

B. 1. $FV_{nc} = 39\,200.00$; $PMT = 3200.00$; $n = 12$; $c = 12$; $m = 12$

0	39 200.00 or (−39 200.00)	3200.00	12			
PV	FV	PMT	N	CPT	%i	0.3740837

$$p = 0.3740837\%$$
$$(1 + i)^{12} = 1.003740837$$
$$1 + i = 1.003740837^{\frac{1}{12}} = 1.000311203$$
$$i = 0.0311203\%$$

The nominal annual rate is $0.0311203\% \times 12 = 0.37344\%$.

3. $PV_{nc} = 62\,400.00$; $PMT = 2600.00$; $n = 50$; $c = \dfrac{1}{2}$; $m = 1$

0	62 400.00	2600.00	50			
FV	PV	PMT	N	CPT	%i	3.3736088

$$p = 3.3736088\%$$
$$(1 + i)^{\frac{1}{2}} = 1.033736088$$
$$1 + i = 1.033736088^2$$
$$1 + i = 1.0686103$$
$$i = 6.86103\%$$

The nominal annual rate is 6.861%.

5. $FV_n(\text{due}) = 6400.00$; $PMT = 200.00$; $n = 18$; $c = \dfrac{12}{2} = 6$; $m = 12$

0	6400.00 or (−6400.00)	200.00	18						
PV	FV	PMT	N	DUE	(or	2nd	CPT)	%i	5.7743238

$$p = 5.77432\%$$
$$(1 + i)^6 = 1.0577432$$
$$1 + i = 1.0577432^{\frac{1}{6}}$$
$$1 + i = 1.00940$$
$$i = 0.94\%$$

The nominal annual rate is $0.94\% \times 12 = 11.28\%$.

7. $PV_{nc}(\text{due}) = 7500.00$; $PMT = 420.00$; $n = 20$; $c = \dfrac{12}{4} = 3$; $m = 12$

0	7500.00	420.00	20						1.2263807
FV	PV	PMT	N	DUE	(or	2nd	CPT	) %i	

$$p = 1.22638\%$$
$$(1 + i)^3 = 1.0122638$$
$$1 + i = 1.0122638^{\frac{1}{3}}$$
$$1 + i = 1.0040713$$
$$i = 0.40713\%$$

The nominal annual rate is $0.40713\% \times 12 = 4.89\%$.

C. 1. $FV_n = 12\,239.76$; $PMT = 350.00$; $n = 24$; $m = 4$

0	12 239.76 or (−12 239.76)	350.00	24			3.125002
PV	FV	PMT	N	CPT	%i	

$i = 3.125002\%$
The nominal annual rate is 12.5%.

3. $FV_n = 13\,600.00$; $PMT = 250.00$; $n = 40$; $m = 4$

0	13 600.00 or (−13 600.00)	250.00	40			1.51149
PV	FV	PMT	N	CPT	%i	

$i = 1.511899\%$
The nominal annual rate is 6.05%.

5. $PV_n = 0.90(11\,400.00) = 10\,260.00$; $PMT = 286.21$; $n = 42$; $c = 12$

0	10 260.00	286.21	42			0.7590734
FV	PV	PMT	N	CPT	%i	

$$i = 0.7590734\%$$

The effective annual rate $f = 1.007590734^{12} - 1$
$$= 1.0949896 - 1$$
$$= 0.0949896$$
$$= 9.50\%$$

7.　$FV_n(\text{due}) = 14\,559.00$;　$PMT = 250.00$;　$n = 40$;　$m = 4$

0	14 559.00 or (−14 559.00)	250.00	40					
PV	FV	PMT	N	DUE	(or	2nd	CPT)	%i	1.7500206

$i = 1.75\%$

The nominal annual rate is $1.75\% \times 4 = 7.0\%$.

9.　$PV_n(\text{due}) = 13\,500.00$;　$PMT = 1200.00$;　$n = 14$;　$c = 2$

0	13 500.00	1200.00	14					
FV	PV	PMT	N	DUE	(or	2nd	CPT)	%i	3.5815675

$i = 3.5815675\%$

The effective annual rate

$f = 1.035815675^2 - 1 = 1.0729141 - 1 = 7.29\%$

11.　$FV_{nc} = 20\,000.00$;　$PMT = 400.00$;　$n = 32$;　$c = \dfrac{12}{4} = 3$;　$m = 12$

0	20 000.00 or (−20 000.00)	400.00	32			
PV	FV	PMT	N	CPT	%i	2.7188829

$p = 2.7189\%$

$(1 + i)^3 = 1.027189$

$1 + i = 1.027189^{\frac{1}{3}}$

$1 + i = 1.0089821$

$i = 0.89821\%$

The nominal annual rate is $0.89821\% \times 12 = 10.778\%$.

13.　$PV_{nc} = 27\,500.00$;　$PMT = 280.00$;　$n = 180$;　$c = \dfrac{2}{12} = \dfrac{1}{6}$;　$m = 2$

0	27 500.00	280.00	180			
FV	PV	PMT	N	CPT	%i	0.755479

$p = 0.7555\%$

$(1 + i)^{\frac{1}{6}} = 1.007555$

$1 + i = 1.007555^6$

$1 + i = 1.0461948$

$i = 4.61948\%$

The nominal annual rate is 9.24%.

15. $PV_{nc}(\text{due}) = 21\,600.00$; $PMT = 680.00$; $n = 36$; $c = \dfrac{1}{12}$; $m = 1$

0	21 600.00	680.00	36				
FV	PV	PMT	N	DUE (or 2nd CPT) %i			0.7344005

$$p = 0.7344\%$$
$$(1 + i)^{\frac{1}{12}} = 1.007344$$
$$1 + i = 1.007344^{12}$$
$$1 + i = 1.0917763$$
$$i = 9.17763\%$$

The nominal annual rate is 9.18% compounded annually.

Review Exercise

1. $FV_n = 10\,000.00$; $i = 2\%$; $n = 20$

   (a) $$10\,000.00 = PMT\left[\frac{1.02^{20} - 1}{0.02}\right]$$
   $$10\,000.00 = 24.2973698 PMT$$
   $$PMT = \$411.57$$

   (b) $$10\,000.00 = PMT\left[\frac{1.02^{20} - 1}{0.02}\right](1.02)$$
   $$10\,000.00 = (24.2973698)(1.02)PMT$$
   $$PMT = \$403.50$$

3. (a) $FV_{nc}(\text{due}) = 32\,000.00$; $i = 3.5\%$; $n = 240$; $c = \dfrac{2}{12} = \dfrac{1}{6}$

   $$p = 1.035^{\frac{1}{6}} - 1 = 1.00575004 - 1 = 0.575004\%$$
   $$32\,000.00 = PMT(1.00575004)\left[\frac{1.00575004^{240} - 1}{0.00575004}\right]$$
   $$32\,000.00 = PMT(1.00575004)(514.6503669)$$
   $$PMT = \$61.82$$

(b) $FV_{nc} = 32\,000.00$; $i = 3.5\%$; $n = 15$; $c = 2$

$p = 1.035^2 - 1 = 1.071225 - 1 = 7.1225\%$

$$32\,000.00 = PMT\left[\frac{1.071225^{15} - 1}{0.071225}\right]$$

$32\,000.00 = 25.3674090 PMT$

$PMT = \$1261.46$

5. $FV_n = 10\,000.00$; $PMT = 300.00$; $i = 2.25\%$

$$10\,000.00 = 300.00\left[\frac{1.0225^n - 1}{0.0225}\right]$$

$$33.333333 = \frac{1.0225^n - 1}{0.0225}$$

$1.0225^n = 1.75$

$n \ln 1.0225 = \ln 1.75$

$0.0222506n = 0.5596158$

$n = 25.15059$ half - years

$= 12.575297$ years

$= 12$ years 7 months

7. $FV_n(\text{due}) = 18\,000.00$; $PMT = 125.00$; $i = 0.5\%$

$$18\,000.00 = 125.00\left[\frac{1.005^n - 1}{0.005}\right](1.005)$$

$$143.28358 = \frac{1.005^n - 1}{0.005}$$

$1.005^n = 1.7164179$

$n \ln 1.005 = \ln 1.7164179$

$0.0049875n = 0.5402395$

$n = 108.31870$

Deposits are required for 109 months.

9. (a) $FV_{nc} = 20\,000.00$; $PMT = 450.00$; $i = 0.5\%$; $c = \dfrac{12}{4} = 3$

$p = 1.005^3 - 1 = 1.0150751 - 1 = 1.50751\%$

$$20\,000.00 = 450.00\left[\frac{1.0150751^n - 1}{0.0150751}\right]$$

$$44.4444444 = \frac{1.0150751^n - 1}{0.0150751}$$

$1.0150751^n = 1.6700044$

$n \ln 1.0150751 = \ln 1.67000444$

$0.01496260n = 0.51282629$

$n = 34.2739$ quarters

(b) $FV_{nc}(\text{due}) = 20\,000.00$; $PMT = 450.00$; $i = 0.5\%$; $c = \dfrac{12}{2} = 6$

$p = 1.005^6 - 1 = 1.0303775 - 1 = 3.03775\%$

$$20\,000.00 = 450.00(1.0303775)\left[\frac{1.0303775^n - 1}{0.0303775}\right]$$

$$43.1341372 = \frac{1.0303775^n - 1}{0.0303775}$$

$1.0303775^n = 2.3103072$

$n \ln 1.0303775 = \ln 2.3103072$

$0.02992524n = 0.8373805$

$n = 27.9824$ half-years

11. $PV_n = 5600.00$; $PMT = 121.85$; $n = 54$; $m = 12$

0	5600.00	121.85	54			
FV	PV	PMT	N	CPT	%i	0.604205

$i = 0.604205\%$

The nominal annual rate is $0.604205\% \times 12 = 7.25\%$.

13. $PV_n(\text{due}) = 9600.00$; $PMT = 235.00$; $n = 48$; $c = 12$

0	9600.00	235.00	48					
FV	PV	PMT	N	DUE	(or	2nd	CPT)	%i	0.7091187

$i = 0.7091187\%$

The effective annual rate $f = 1.007091187^{12} - 1 = 1.0884928 - 1 = 8.85\%$

15. Amount after 5 years: $PV = 10\,000.00$; $i = 3.75\%$; $n = 10$

$FV = 10\,000.00\left(1.0375^{10}\right) = 10\,000.00(1.4450439) = \$14\,450.44$

Annuity: $PV_n = 14\,450.44$; $PMT = 2000.00$; $i = 3.75\%$

$$14\,450.44 = 2000.00\left[\frac{1 - 1.0375^{-n}}{0.0375}\right]$$

$$7.2252197 = \frac{1 - 1.0375^{-n}}{0.0375}$$

$$1.0375^{-n} = 0.7290543$$

$$-n \ln 1.0375 = \ln 0.7290543$$

$$-0.0368140n = -0.3160071$$

$$n = 8.5838838 \text{ half - years}$$

The term consists of 9 semi-annual payments.

17. $FV_{nc}(\text{due}) = 8400.00$; $PMT = 400.00$; $n = 16$; $c = \dfrac{4}{2} = 2$; $m = 4$

0	8400.00 or	400.00	16
	(−8400.00)		

| PV | | FV | | PMT | | N | | DUE | (or | 2nd | | CPT |) | %i | | 3.1287658 |

$$p = 3.128766\%$$

$$(1 + i)^2 = 1.03128766$$

$$1 + i = 1.03128766^{\frac{1}{2}}$$

$$1 + i = 1.0155233$$

$$i = 1.55233\%$$

The nominal annual rate is 6.21%.

19. $FV_n(\text{due}) = 180\,000.00\left(1.075^{-12}\right) = 180\,000.00(0.4198541) = \$75\,573.74$

$PMT = 1000.00$; $i = 7.5\%$

$$75\,573.74 = 1000.00(1.075)\left[\frac{1.075^n - 1}{0.075}\right]$$

$$70.301153 = \frac{1.075^n - 1}{0.075}$$

$$1.075^n = 6.2725865$$

$$n \ln 1.075 = \ln 6.2725865$$

$$0.0723207n = 1.8361888$$

$$n = 25.389533 \text{ years}$$

Deposits are required for 26 years.

21. Amount after 15 years:
    PMT = 125.00; $i = 0.625\%$; $n = 180$

    $FV = 125.00\left(1.00625^{180} - 1\right) = 125.00(1.1122763) = \$41\,389.03$

    Annuity rate:
    $PV_n(\text{due}) = 41\,389.03$; PMT = 1200.00; $n = 48$; $c = 4$

0	41 389.03	1200.00	48			
FV	PV	PMT	N	CPT	%i	1.4388887

    The effective annual rate $f = 1.01438889^4 - 1 = 1.0588097 - 1 = 5.881\%$

23. Amount at age 65:
    PV = 25 000.00; $i = 2.75\%$; $n = 60$

    $FV = 25\,000.00\left(1.0275^{60}\right) = 25\,000.00(5.0922514) = \$127\,306.28$

    Term of annuity:
    $PV_n = 127\,306.28$; PMT = 6000.00; $i = 1.5\%$

    $$127\,306.28 = 6000.00\left[\frac{1 - 1.015^{-n}}{0.015}\right]$$

    $$21.217713 = \frac{1 - 1.015^{-n}}{0.015}$$

    $$1.015^{-n} = 0.6817343$$
    $$-n\ln 1.015 = \ln 0.6817343$$
    $$-0.0148886n = -0.3831153$$
    $$n = 25.7321 \text{ quarters}$$

    The annuity will run for 26 quarters (6 years 6 months).

25. (a) PMT = 850.00; $i = 0.875\%$; $n = 240$

    $$PV_n(\text{due}) = 850.00(1.00875)\left[\frac{1 - 1.00875^{-240}}{0.00875}\right]$$
    $$= 850.00(1.00875)(100.16227)$$
    $$= \$85\,882.89$$

    (b) Interest $= 240(850.00) - 85\,882.89 = 204\,000.00 - 85\,882.89 = \$118\,117.11$

    (c) $FV_n(\text{due}) = 85\,882.89$; $i = 0.875\%$; $n = 180$

    $$85\,882.89 = PMT(1.00875)\left[\frac{1.00875^{180} - 1}{0.00875}\right]$$
    $$85\,882.89 = PMT(1.00875)(434.0298)$$
    $$PMT = \$196.16$$

    (d) Total interest $= 204\,000.00 - 180(196.16)$
    $$= 204\,000.00 - 35\,308.80 = \$168\,691.20$$

27. $PV = 75\,000.00$; $i = 4\%$; $n = 20$

$FV = 75\,000.00\left(1.04^{20}\right) = 75\,000.00(2.1911231) = \$164\,334.23$

$FV_{nc}(\text{due}) = 164\,334.23$; $i = 0.625\%$; $n = 40$; $c = \dfrac{12}{4} = 3$

$p = 1.00625^3 - 1 = 1.0188674 - 1 = 1.88674\%$

$164\,334.23 = PMT(1.0188674)\left[\dfrac{1.0188674^{40} - 1}{0.0188674}\right]$

$164\,334.23 = PMT(1.0188674)(58.940925)$

$\quad PMT = \$2736.49$

29. $PV_{nc}(\text{defer}) = 200\,000.00$; $n = 80$; $d = 20$; $i = 10\%$; $c = \dfrac{1}{4}$

$p = 1.10^{\frac{1}{4}} - 1 = 1.02411369 - 1 = 2.411369\%$

$200\,000.00 = PV_{nc}(\text{due})\left(1.02411369^{-20}\right)$

$200\,000.00 = 0.6209213\,PV_{nc}(\text{due})$

$PV_{nc}(\text{due}) = 322\,102.00$

$322\,102.00 = PMT(\text{due})(1.02411369)\left[\dfrac{1 - 1.0241137^{-80}}{0.0241137}\right]$

$322\,102.00 = PMT(\text{due})(1.02411369)(35.305926)$

$PMT(\text{due}) = \$8908.36$

31. Amount at retirement:
$PMT = 500.00$; $i = 2.25\%$; $n = 100$

$FV_n = 500.00\left[\dfrac{1.0225^{100} - 1}{0.0225}\right] = 500.00(366.8465023) = \$183\,423.25$

Quarterly payment:
$PV_n = 183\,423.25$; $i = 2.25\%$; $n = 80$

$183\,423.25 = PMT\left[\dfrac{1 - 1.0225^{-80}}{0.0225}\right]$

$183\,423.25 = 36.9497808\,PMT$

$\quad PMT = \$4964.12$

33. $PV_n(\text{defer}) = 75\,000.00; \; PMT = 6000.00; \; d = 20 - 1 = 19; \; i = 2.125\%$

$$75\,000.00 = 6000.00\left(1.02125^{-19}\right)\left[\frac{1 - 1.02125^{-n}}{0.02125}\right]$$

$$75\,000.00 = 6000.00(0.6706419)\left[\frac{1 - 1.02125^{-n}}{0.02125}\right]$$

$$0.3960758 = 1 - 1.02125^{-n}$$

$$1.02125^{-n} = 0.6039242$$

$$-n \ln 1.02125 = \ln 0.6039242$$

$$-0.02102737n = -0.5043066$$

$$n = 23.98334 \text{ quarters}$$

24 quarterly payments have to be made.

35. $PV = 37\,625.00; \; i = 1.75\%; \; d = 3; \; c = 4$

$p = 1.0175^4 - 1 = 1.071859 - 1 = 7.1859\%$

$$37\,625.00 = \frac{PMT}{0.071859}\left(1.071859^{-3}\right)$$

$$2703.6949 = PMT(0.8120580)$$

$$PMT = \$3329.44$$

37. $PMT = 1350.00; \; FV_{nc}(\text{due}) = 125\,000.00; \; i = 1.75\%; \; c = 4$

$p = 1.0175^4 - 1 = 1.0718590 - 1 = 7.18590\%$

$$125\,000.00 = 1350.00(1.071859)\left[\frac{1.071859^n - 1}{0.071859}\right]$$

$$86.385049 = \frac{1.071859^n - 1}{0.071859}$$

$$1.071859^n = 7.2075433$$

$$n \ln 1.071859 = \ln 7.2075433$$

$$0.0693945n = 1.9751282$$

$$n = 28.4623 \text{ years}$$

Contributions are needed for 29 years.

CHAPTER 13

39. $PV_{nc}(\text{due}) = 50\,000.00$; $PMT = 5200.00$; $i = 0.75\%$; $c = \dfrac{12}{2} = 6$

$p = 1.0075^6 - 1 = 1.0458522 - 1 = 4.58522\%$

$$50\,000.00 = 5200.00(1.0458522)\left[\frac{1 - 1.0458522^{-n}}{0.0458522}\right]$$

$$9.1938274 = \frac{1 - 1.0458522^{-n}}{0.0458522}$$

$$0.4215572 = 1 - 1.0458522^{-n}$$

$$1.0458522^{-n} = 0.5784428$$

$$-n\ln 1.0458522 = \ln 0.5784428$$

$$-0.0448321n = -0.5474156$$

$$n = 12.21035 \text{ half - years}$$

The term is 13 semi-annual periods (6.5 years).

41. Amount after 20 years:
$PMT = 900.00$; $i = 2.75\%$; $n = 40$

$$PV_n = 900.00\left[\frac{1.0275^{40} - 1}{0.0275}\right] = 900.00(71.2681450) = \$64\,141.33$$

Amount 5 years later:
$PV = 64\,141.33$; $i = 2.75\%$; $n = 10$

$$FV = 64\,141.33\left(1.0275^{10}\right) = 64\,141.33(1.3116510) = \$84\,131.04$$

Annuity payment:
$PV_n = 84\,131.04$; $i = 0.5\%$; $n = 180$

$$84\,131.04 = PMT\left[\frac{1 - 1.005^{-180}}{0.005}\right]$$

$$84\,131.04 = 118.5035147PMT$$

$$PMT = \$709.95$$

43. (a) $PMT = 400.00$; $FV_n = 92\,000.00$; $n = 80$; $c = 4$

0	92 000.00 or (−92 000.00)	400.00	80		
PV	FV	PMT	N	CPT	%i	2.3472272

$i = 2.3472272\%$

The effective annual rate $f = 1.023472272^4 - 1 = 1.0972468 - 1 = 9.725\%$

222

(b) $PV_n = 92\,000.00; \ PMT = 1350.00$

Nominal rate $= 2.347\% \times 4 = 9.39\%; \ i = 0.7825\%$

$$92\,000.00 = 1350.00\left[\frac{1 - 1.007825^{-n}}{0.007825}\right]$$

$$68.148148 = \frac{1 - 1.007825^{-n}}{0.007825}$$

$$1.007825^{-n} = 0.4667407$$

$$-n \ln 1.007825 = \ln 0.4667407$$

$$-0.0077945435n = -0.761981422$$

$$n = 97.758313$$

The term is 98 months.

45.   Amount after 16 years:

$PMT = 50.00; \ i = 0.5\%; \ n = 192$

$$FV_n(\text{due}) = 50.00(1.005)\left[\frac{1.005^{192} - 1}{0.005}\right] = 50.00(1.005)(321.09134) = \$16\,134.84$$

Number of withdrawals:

$PV_n(\text{due}) = 16\,134.84; \ PMT = 375.00; \ i = 0.5\%$

$$16\,134.84 = 375.00\left[\frac{1 - 1.005^{-n}}{0.005}\right](1.005)$$

$$42.812179 = \frac{1 - 1.005^{-n}}{0.005}$$

$$1.005^{-n} = 0.7859391$$

$$-n \ln 1.005 = \ln 0.7859391$$

$$-0.0049875n = -0.24087597$$

$$n = 48.295933 \text{ months}$$

Withdrawals can be made for 49 months.

47. $PV_{nc}(\text{defer}) = 25\,000.00; \ PMT(\text{due}) = 1400.00; \ i = 5.75\%; \ d = 4; \ c = \dfrac{2}{4} = \dfrac{1}{2}$

$p = 1.0575^{\frac{1}{2}} - 1 = 1.0283482 - 1 = 2.83482\%$

$$25\,000.00 = PV_{nc}(\text{due})\left(1.0283482^{-4}\right)$$

$$25\,000.00 = 0.8942094 PV_{nc}(\text{due})$$

$$PV_{nc}(\text{due}) = 27\,957.657$$

$$27\,957.657 = 1400.00(1.0283482)\left[\frac{1 - 1.0283482^{-n}}{0.0283482}\right]$$

$$0.5505009 = 1 - 1.0283482^{-n}$$

$$1.0283482^{-n} = 0.4494991$$

$$-n \ln 1.0283482 = \ln 0.4494991$$

$$-0.027953826n = -0.7996214$$

$$n = 28.605079 \text{ quarters}$$

1. $\text{PMT} = 4800.00; \ i = 1.5\%; \ n = 24$

    $PV_n = 4800.00\left[\dfrac{1 - 1.015^{-24}}{0.015}\right] = 4800.00(20.0304054) = \$96\,145.95$

    $FV_n = 96\,145.95; \ i = 0.625\%; \ n = 120$

    $96\,145.95 = \text{PMT}\left[\dfrac{1.00625^{120} - 1}{0.00625}\right]$

    $96\,145.95 = 177.9303419\text{PMT}$

    $\text{PMT} = \$540.36$

3. $PV_n = 14\,400.00; \ \text{PMT} = 600.00; \ i = 2.625\%$

    $14\,400.00 = 600.00\left[\dfrac{1 - 1.02625^{-n}}{0.02625}\right]$

    $24 = \dfrac{1 - 1.02625^{-n}}{0.02625}$

    $0.63 = 1 - 1.02625^{-n}$

    $1.02625^{-n} = 0.37$

    $-n\ln 1.02625 = \ln 0.37$

    $-0.0259114n = -0.9942523$

    $n = 38.37 \text{ quarterly payments}$

5. $\text{PMT} = 3000.00; \ i = 2.5\%; \ n = 120$

    $PV_n(\text{due}) = 3000.00(1.025)\left[\dfrac{1 - 1.025^{-120}}{0.025}\right]$

    $\qquad = 3000.00(1.025)(37.933687)$

    $\qquad = 113\,801.06(1.025) = \$116\,646.09$

    $FV_{nc}(\text{due}) = 116\,646.09; \ i = 2.5\%; \ n = 240; \ c = \dfrac{4}{12} = \dfrac{1}{3}$

    $p = 1.025^{0.3333333} - 1 = 1.008264836 - 1 = 0.8264836\%$

    $116\,646.09 = \text{PMT}(1.0082648)\left[\dfrac{1.008264836^{240} - 1}{0.008264836}\right]$

    $116\,646.09 = \text{PMT}(1.0082648)(751.32345)$

    $116\,646.09 = 757.53298\text{PMT}$

    $\text{PMT} = \$153.98$

7. $PV = 57\,426.00$; $i = 0.5\%$; $n = 72$

$FV = 57\,426.00(1.005)^{72} = 57\,426.00(1.4320443) = \$82\,236.58$

$PV_n(\text{due}) = 82\,236.58$; $i = 1.475\%$; $PMT = 3600.00$

$$82\,236.58 = 3600.00(1.01475)\left[\frac{1 - 1.01475^{-n}}{0.01475}\right]$$

$$22.511451 = \frac{1 - 1.01475^{-n}}{0.01475}$$

$$0.3320439 = 1 - 1.01475^{-n}$$

$$1.01475^{-n} = 0.6679561$$

$$-n\ln 1.01475 = \ln 0.6679561$$

$$-0.0146423n = -0.4035328$$

$$n = 27.5594 \text{ quarters}$$

The term is 28 quarters, i.e., 84 months.

9. Let the size of the equal payments be $\$x$.; the focal date is 4 years from now; $i = 2.375\%$.

$$\begin{aligned} 5000.00 &= x(1.02375)^{16} + x(1.02375)^{11} + x(1.02375)^{7} + x(1.02375)^{4} \\ &= 1.4558031x + 1.2945924x + 1.1785755x + 1.0984383x \\ &= 5.0274093x \\ x &= \$994.55 \end{aligned}$$

11. $FV_n(\text{due}) = 40\,000.00$; $PMT = 1400.00$; $n = 24$; $c = 4$

0	40 000.00 or	1400.00	24					
	(−40 000.00)							
PV	FV	PMT	N	DUE	(or	2nd	CPT)	%i	1.3687196

$i = 1.3687196\%$

The effective annual rate $f = 1.013687196^{4} - 1 = 1.0558831 - 1 = 5.588\%$

13. $FV_{nc} = 20\,000.00$; $PMT = 491.00$; $i = 0.5\%$; $c = \dfrac{12}{4} = 3$

$p = 1.005^{3} - 1 = 1.0150751 - 1 = 1.50751\%$

$$20\,000.00 = 491.00\left[\frac{1.0150751^{n} - 1}{0.0150751}\right]$$

$$40.7331976 = \frac{1.0150751^{n} - 1}{0.0150751}$$

$$0.6140570 = 1.0150751^{n} - 1$$

$$n\ln 1.0150751 = \ln 1.6140570$$

$$0.0149626n = 0.4787509$$

$$n = 31.9965 \text{ quarters}$$

15. PMT $= 800.00$; $i = 1.875\%$; $n = 20$

$$FV_n = 800.00\left[\frac{1.01875^{20} - 1}{0.01875}\right] = 800.00(23.997228) = \$19\,197.78$$

PV $= 19\,197.78$; $i = 1.875\%$; $n = 24$

FV $= 19\,197.78(1.01875)^{24} = 19\,197.78(1.5617910) = \$29\,982.92$

$PV_n = 29\,982.92$; $i = 0.45\%$; $n = 360$

$$29\,982.92 = PMT\left[\frac{1 - 1.0045^{-360}}{0.0045}\right]$$

$29\,982.92 = 178.08462\,PMT$

    PMT $= \$168.36$

17. PMT $= 1680.00$; $i = 2.75\%$; $n = 16$

$$FV_n(\text{due}) = 1680.00(1.0275)\left[\frac{1.0275^{16} - 1}{0.0275}\right]$$

$$= 1680.00(1.0275)(19.7639795)$$

$$= 33\,203.49(1.0275)$$

$$= \$34\,116.58$$

PV $= 34\,116.58$; $i = 2.75\%$; $n = 14$

FV $= 34\,116.58(1.0275)^{14} = 34\,116.58(1.4619941) = \$49\,878.24$

$PV_n = 49\,878.24$; $i = 0.5\%$; $n = 240$

$$49\,878.24 = PMT\left[\frac{1 - 1.005^{-240}}{0.005}\right]$$

$49\,878.24 = 139.58077\,PMT$

    PMT $= \$357.34$

14 Amortization of debts

Exercise 14.1

A. 1. (a) $PV_n = 12\,000.00$; $i = 2.5\%$; $n = 32$

$$12\,000.00 = PMT\left[\frac{1 - 1.025^{-32}}{0.025}\right]$$

$12\,000.00 = 21.849178\,PMT$
$ PMT = \$549.22$

(b) $PMT = 549.22$; $i = 2.5\%$; $n = 32 - 20 = 12$

$$PV_{20} = 549.22\left[\frac{1 - 1.025^{-12}}{0.025}\right]$$

$\phantom{PV_{20}} = 549.22(10.2577646)$
$\phantom{PV_{20}} = \$5633.77$

(c) Interest in $PMT_{21} = 5633.77(0.025) = \$140.84$

(d) Principal repaid $= 549.22 - 140.84 = \$408.38$

3. (a) $PV_n = 15\,000.00$; $i = 4\%$; $n = 20$

$$15\,000.00 = PMT\left[\frac{1 - 1.04^{-20}}{0.04}\right]$$

$15\,000.00 = 13.5903263\,PMT$
$ PMT = \$1103.73$

(b) $PMT = 1103.73$; $i = 4\%$; $n = 20 - 15 = 5$

$$PV_{15} = 1103.73\left[\frac{1 - 1.04^{-5}}{0.04}\right]$$

$\phantom{PV_{15}} = 1103.73(4.4518223)$
$\phantom{PV_{15}} = \$4913.61$

(c) Interest in $PMT_{16} = 4913.61(0.04) = \$196.54$

(d) Principal repaid $= 1103.73 - 196.54 = \$907.19$

5.   (a) $PV_n = 5500.00;\ i = 0.75\%;\ n = 48$

$$PV_{14} = 136.87\left[\frac{1 - 1.0075^{-34}}{0.0075}\right]$$
$$= 136.87(29.9127762)$$
$$= \$4094.16$$

    (b) $PMT = 136.87;\ i = 0.75\%;\ n = 48 - 14 = 34$
$$PV_{14} = 136.87\left[\frac{1 - 1.0075^{-34}}{0.0075}\right]$$
$$= 136.87(29.9127762)$$
$$= \$4094.16$$

    (c)   Interest paid in $PMT_{15} = 4094.16(0.0075) = \$30.71$

    (d)   Principal repaid $= 136.87 - 30.71 = \$106.16$

B.   1.   (a) $PV_n = 12\,000.00;\ PMT = 750.00;\ i = 2.0\%$

$$12\,000.00 = 750.00\left[\frac{1 - 1.02^{-n}}{0.02}\right]$$
$$16 = \frac{1 - 1.02^{-n}}{0.02}$$
$$0.32 = 1 - 1.02^{-n}$$
$$1.02^{-n} = 0.68$$
$$-n\ln 1.02 = \ln 0.68$$
$$-0.01980263n = -0.3856625$$
$$n = 19.47532$$

19 payments of $750.00 plus a final payment less than $750.00.

    (b)   Using the retrospective method:
$PV = 12\,000.00;\ i = 2.0\%;\ n = 16;\ PMT = 750.00$

$$FV = 12\,000.00\left(1.02^{16}\right) = 12\,000.00(1.3727857) = \$16\,473.43$$

$$FV_{16} = 750.00\left[\frac{1.02^{16} - 1}{0.02}\right] = 750.00(18.6392853) = \$13\,979.46$$

$$PV_{16} = 16\,473.43 - 13\,979.46 = \$2493.97$$

3. (a) $PV_n = 21\,000.00;\ PMT = 2000.00;\ i = 4.5\%$

$$21\,000.00 = 2000.00\left[\frac{1 - 1.045^{-n}}{0.045}\right]$$

$$10.5 = \frac{1 - 1.045^{-n}}{0.045}$$

$$0.4725 = 1 - 1.045^{-n}$$

$$1.045^{-n} = 0.5275$$

$$-n\ln 1.045 = \ln 0.5275$$

$$-0.044016885n = -0.6396064$$

$$n = 14.530933$$

14 payments of $2000.00 plus a final payment of less than $2000.00.

(b) Using the retrospective method:

$$FV. = 21\,000.00\left(1.045^{10}\right) = 21\,000.00(1.5529694) = \$32\,612.36$$

$$FV_{10} = 2000.00\left[\frac{1.045^{10} - 1}{0.045}\right] = 2000.00(12.288209) = \$24\,576.42$$

$$PV_{10} = 32\,612.36 - 24\,576.42 = \$8035.94$$

C. 1. (a) $PV_n = 36\,000.00 - 4000.00 = 32\,000.00;\ i = 2\%;\ n = 60$

$$32\,000.00 = PMT\left[\frac{1 - 1.02^{-60}}{0.02}\right]$$

$$32\,000.00 = 34.760887PMT$$

$$PMT = \$920.57$$

(b) $PMT = 920.57;\ i = 2\%;\ n = 60 - 40 = 20$

$$PV_{40} = 920.57\left[\frac{1 - 1.02^{-20}}{0.02}\right]$$

$$= 920.57(16.351433)$$

$$= \$15\,052.64$$

(c) Total paid $= 4000.00 + 60(920.57) = \$59\,234.20$

(d) Total interest paid $= 55\,234.20 - 32\,000.00 = \$23\,234.20$

3. $PV_n = 10\,000.00;\ i = 10\%;\ n = 7$

$$10\,000.00 = PMT\left[\frac{1 - 1.10^{-7}}{0.10}\right]$$

$$10\,000.00 = 4.868419 PMT$$

$$PMT = \$2054.05$$

Amortization Schedule

Payment number	Amount paid	Interest paid	Principal repaid	Outstanding principal
0				10 000.00
1	2054.05	1000.00	1054.05	8945.95
2	2054.05	894.60	1159.45	7786.50
3	2054.05	778.65	1275.40	6511.10
4	2054.05	651.11	1402.94	5108.16
5	2054.05	510.82	1543.23	3564.93
6	2054.05	356.49	1697.56	1867.37
7	2054.11	186.74	1867.37	—
Total	14 378.41	4378.41	10 000.00	

5. $PV_n = 9200.00;\ PMT = 2000.00;\ i = 13\%$

Amortization Schedule

Payment number	Amount paid	Interest paid	Principal repaid	Outstanding principal
0				9200.00
1	2000.00	1196.00	804.00	8396.00
2	2000.00	1091.48	908.52	7487.48
3	2000.00	973.37	1026.63	6460.85
4	2000.00	839.91	1160.09	5300.76
5	2000.00	689.10	1310.90	3989.86
6	2000.00	518.68	1481.32	2508.54
7	2000.00	326.11	1673.89	834.65
8	943.16	108.51	834.65	—
Total	14 943.16	5743.16	9200.00	

7. $PMT = 2054.05;\ i = 10\%;\ n = 7 - 3 = 4$

$$PV_3 = 2054.05\left[\frac{1 - 1.10^{-4}}{0.10}\right] = 2054.05(3.1698654) = \$6511.06$$

Interest in $PMT_4 = 6511.06(0.10) = \$651.11$

9. Use the retrospective method:
    $PV_n = 9200.00$; $PMT = 2000.00$; $i = 13\%$
    Accumulated value of principal after 3 payments

    $= 9200.00\left(1.13^3\right) = 9200.00(1.442897) = \$13\,274.65$

    Accumulated value of first 3 payments

    $= 2000.00\left[\dfrac{1.13^3 - 1}{0.13}\right] = 2000.00(3.4069) = \$6813.80$

    $PV_3 = 13\,274.65 - 6813.80 = \$6460.85$

    Interest paid in $PMT_4 = 6460.85(0.13) = \$839.91$

    Principal repaid $= 2000.00 - 839.91 = \$1160.09$

11. (a) $PV_n = 85\,000.00$; $i = 2\%$; $n = 32$

    $$85\,000.00 = PMT\left[\dfrac{1 - 1.02^{-32}}{0.02}\right]$$

    $85\,000.00 = 23.468335PMT$

    $PMT = \$3621.90$

    (b) $PMT = 3621.90$; $i = 2\%$; $n = 32 - 15 = 17$

    $PV_{15} = 3621.90\left[\dfrac{1 - 1.02^{-17}}{0.02}\right]$

    $= 3621.90(14.291872)$

    $= \$51\,763.73$

    Interest paid in $PMT_{16} = 51\,763.73(0.02) = \$1035.27$

    (c) $PMT = 3621.90$; $i = 2\%$; $n = 32 - 19 = 13$

    $PV_{19} = 3621.90\left[\dfrac{1 - 1.02^{-13}}{0.02}\right]$

    $= 3621.90(11.348374)$

    $= \$41\,102.67$

    Interest paid in $PMT_{20} = 41\,102.67(0.02) = \$822.05$

    Principal repaid $= 3621.90 - 822.05 = \$2799.85$

(d)   Last 3 payments are $PMT_{30}, PMT_{31}, PMT_{32}$.

PMT = 3621.90; i = 2%; n = 32 − 29 = 3

$$PV_{29} = 3621.90 \left[\frac{1 - 1.02^{-3}}{0.02} \right] = 3621.90(2.8838833) = \$10\,445.14$$

Partial Amortization Schedule

Payment number	Amount paid	Interest paid	Principal repaid	Outstanding balance
0				85 000.00
1	3621.90	1700.00	1921.90	83 078.10
2	3621.90	1661.56	1960.34	81 117.76
3	3621.90	1622.36	1999.54	79 118.22
⋮	⋮	⋮	⋮	⋮
29	⋮	⋮	⋮	10 445.14
30	3621.90	208.90	3413.00	7032.14
31	3621.90	140.64	3481.26	3550.88
32	3621.90	71.02	3550.88	—
Total	115 900.80	30 900.80	85 000.00	

13.   (a) PV_n = 24 000.00; PMT = 2500.00%; i = 5.5%

$$24\,000.00 = 2500.00 \left[\frac{1 - 1.055^{-n}}{0.055} \right]$$

$$9.60 = \frac{1 - 1.055^{-n}}{0.055}$$

$$0.528 = 1 - 1.055^{-n}$$

$$1.055^{-n} = 0.472$$

$$-n \ln 1.055 = \ln 0.472$$

$$-0.0535408n = -0.7507763$$

$$n = 14.022508$$

14 payments of $2500.00 and a final payment of less than $2500.00 are required.

(b)   Use the retrospective method:

PV_n = 24 000.00; PMT = 2500.00; i = 5.5%; n = 5

$$FV = 24\,000.00\left(1.055^5\right) = 24\,000.00(1.306960) = \$31\,367.04$$

$$FV_5 = 2500.00 \left[\frac{1.055^5 - 1}{0.055} \right] = 2500.00(5.581091) = \$13\,952.73$$

$$PV_5 = 31\,367.04 - 13\,952.73 = \$17\,414.31$$

Interest paid in $PMT_6 = 17\,414.31(0.055) = \$957.79$

(c)   Use the retrospective method:

$PV_n = 24\,000.00$; $PMT = 2500.00$; $i = 5.5\%$; $n = 9$

$FV = 24\,000.00(1.055^9) = 24\,000.00(1.6190943) = \$38\,858.26$

$$FV_9 = 2500.00\left[\frac{1.055^9 - 1}{0.055}\right] = 2500.00(11.2562595) = \$28\,140.65$$

$PV_9 = 38\,858.26 - 28\,140.65 = \$10\,717.61$

Interest paid in $PMT_{10} = 10\,717.61(0.055) = \$589.47$

Principal repaid $= 2500.00 - 589.47 = \$1910.53$

(d)   Last 3 payments are PMT_{13}, PMT_{14}, PMT_{15}.

To find the balance outstanding after 12 payments, use the retrospective method or use the prospective method:

$PMT = 2500.00$; $i = 5.5\%$; $n = 14.022508 - 12 = 2.022508$

$$PV_{12} = 2500.00\left[\frac{1 - 1.055^{-2.022508}}{0.055}\right]$$

$$= 2500.00(1.8659937)$$

$$= \$4664.98$$

Partial Amortization Schedule

Payment number	Amount paid	Interest paid	Principal repaid	Outstanding balance
0				24 000.00
1	2 500.00	1 320.00	1 180.00	22 820.00
2	2 500.00	1 255.10	1 244.90	21 575.10
3	2 500.00	1 186.63	1 313.37	20 261.73
⋮	⋮	⋮	⋮	⋮
⋮	⋮	⋮	⋮	⋮
12	⋮	⋮	⋮	4 664.98
13	2 500.00	256.57	2 243.43	2 421.55
14	2 500.00	133.19	2 366.81	54.74
15	57.75	3.01	54.74	—
Total	35 057.75	11 057.75	24 000.00	

Exercise 14.2

A.   1.   (a) $PV_n = 36\,000.00$; $i = 2\%$; $n = 40$; $c = \dfrac{4}{2} = 2$

$p = 1.02^2 - 1 = 1.0404 - 1 = 4.04\%$

$$36\,000.00 = PMT\left[\frac{1 - 1.0404^{-40}}{0.0404}\right]$$

$36\,000.00 = 19.6755018\,PMT$

$PMT = \$1829.69$

(b) PMT = 1829.69; p = 4.04%; n = 40 − 25 = 15

$$PV_{25} = 1829.69\left[\frac{1 - 1.0404^{-15}}{0.0404}\right]$$

$$= 1829.69(11.0873542)$$

$$= \$20\,286.42$$

(c) Interest in PMT_{26} = 20 286.42(0.0404) = \$819.57

(d) Principal repaid = 1829.69 − 819.57 = \$1010.12

3. (a) PV_{nc} = 8500.00; i = 3%; n = 60; $c = \dfrac{2}{12} = \dfrac{1}{6}$

$p = 1.03^{\frac{1}{6}} - 1 = 1.00493862 - 1 = 0.493862\%$

$$8500.00 = PMT\left[\frac{1 - 1.00493862^{-60}}{0.00493862}\right]$$

$$8500.00 = 51.817308PMT$$

$$PMT = \$164.04$$

(b) PMT = 164.04; p = 0.493862%; n = 60 − 30 = 30

$$PV_{30} = 164.04\left[\frac{1 - 1.00493862^{-30}}{0.00493862}\right]$$

$$= 164.04(27.819644)$$

$$= \$4563.53$$

(c) Interest paid in PMT_{31} = 4563.53(0.00493862) = \$22.54

(d) Principal repaid = 164.04 − 22.54 = \$141.50

5. (a) PV_{nc} = 45 000.00; i = 0.75%; n = 30; $c = \dfrac{12}{2} = 6$

$p = 1.0075^{6} - 1 = 1.0458522 - 1 = 4.58522\%$

$$45\,000.00 = PMT\left[\frac{1 - 1.0458522^{-30}}{0.0458522}\right]$$

$$45\,000.00 = 16.126823PMT$$

$$PMT = \$2790.38$$

(b) PMT = 2790.38; p = 4.58522%; n = 30 − 12 = 18

$$PV_{12} = 2790.38\left[\frac{1 - 1.0458522^{-18}}{0.0458522}\right]$$

$$= 2790.38(12.077830)$$

$$= \$33\,701.74$$

(c) Interest paid in PMT_{13} = 33 701.74(0.0458522) = \$1545.30

(d) Principal repaid = 2790.38 − 1545.30 = \$1245.08

B. 1. (a) PV_{nc} = 6000.00; PMT = 400.00; i = 0.5%; $c = \dfrac{12}{4} = 3$

$p = 1.005^3 - 1 = 1.0150751 - 1 = 1.50751\%$

$$6000.00 = 400.00\left[\frac{1 - 1.0150751^{-n}}{0.0150751}\right]$$

$$15.00 = \frac{1 - 1.0150751^{-n}}{0.0150751}$$

$$0.2261265 = 1 - 1.0150751^{-n}$$

$$1.0150751^{-n} = 0.7738735$$

$$-n\ln 1.0150751 = \ln 0.7738735$$

$$-0.0149626n = -0.2563469$$

$$n = 17.13251$$

Number of payments is 18.

(b) Use the retrospective method:
Accumulated value of principal:
PV = 6000.00; i = 0.5%; n = 10(3) = 30

$FV = 6000.00\left(1.005^{30}\right) = 6000.00(1.1614001) = \$6968.40$

Accumulated value of payments:
PMT = 400.00; p = 1.50751%; n = 10

$$FV_{10} = 400.00\left[\frac{1.0150751^{10} - 1}{0.0150751}\right] = 400.00(10.706383) = \$4282.55$$

PV_{10} = 6968.40 − 4282.55 = \$2685.85

Alternatively, use the prospective method:
PMT = 400.00; p = 1.50751%; n = 17.13251 − 10 = 7.13251

$$PV_{10} = 400.00\left[\frac{1 - 1.0150751^{-7.13251}}{0.0150751}\right]$$

$$= 400.00(6.714616)$$

$$= \$2685.85$$

3. (a) $PV_{nc} = 23\,500.00$; $PMT = 1800.00$; $i = 7\%$; $c = \dfrac{1}{4}$

$$p = 1.07^{\frac{1}{4}} - 1 = 1.017058525 - 1 = 1.7058525\%$$

$$23\,500.00 = 1800.00 \left[\frac{1 - 1.017058525^{-n}}{0.017058525} \right]$$

$$13.055556 = \frac{1 - 1.017058525^{-n}}{0.017058525}$$

$$0.2227085 = 1 - 1.017058525^{-n}$$

$$1.017058525^{-n} = 0.7772915$$

$$-n \ln 1.017058525 = \ln 0.7772915$$

$$-0.01691466n = -0.25193984$$

$$n = 14.894762$$

The number of payments is 15.

(b) Using the retrospective method:
Accumulated value of principal:

$$PV = 23\,500.00; \; i = 7\%; n = 14\left(\frac{1}{4}\right) = 3.50$$

$$FV = 23\,500.00\left(1.07^{3.5}\right) = 23\,500.00(1.2671943) = \$29\,779.07$$

Accumulated value of payments:

$$PMT = 1800.00; \; p = 1.70585\%; \; n = 14$$

$$FV_{14} = 1800.00 \left[\frac{1.0170585^{14} - 1}{0.0170585} \right] = 1800.00(15.663388) = \$28\,194.10$$

$$PV_{14} = 29\,779.07 - 28\,194.10 = \$1584.97$$

Alternatively:

$$PMT = 1800.00; \; p = 1.70585\%; \; n = 14.894762 - 14 = 0.894762$$

$$PV_{14} = 1800.00 \left[\frac{1 - 1.0170585^{-0.894762}}{0.0170585} \right]$$

$$= 1800.00(0.8805360)$$

$$= \$1584.96$$

C. 1. (a) $PV_{nc} = 36\,000.00$; $i = 3.5\%$; $n = 300$; $c = \dfrac{2}{12} = \dfrac{1}{6}$

$$p = 1.035^{\frac{1}{6}} - 1 = 1.0057500 - 1 = 0.575\%$$

$$36\,000.00 = PMT \left[\frac{1 - 1.00575^{-300}}{0.00575} \right]$$

$$36\,000.00 = 142.77296 PMT$$

$$PMT = \$252.15$$

(b) $\text{PMT} = 252.15$; $p = 0.575\%$; $n = 300 - 36 = 264$

$$\text{PV}_{36} = 252.15\left[\frac{1 - 1.00575^{-264}}{0.00575}\right]$$

$$= 252.15(135.63398)$$

$$= \$34\,200.11$$

(c) Total paid $= 252.15(36) = \$9077.40$
Reduction in principal $= 36\,000.00 - 34\,200.11 = \$1799.89$
Interest paid $= 9077.40 - 1799.89 = \$7277.51$

(d) $\text{PV}_{nc} = 34\,200.11$; $i = 4.5\%$; $n = 264$; $c = \dfrac{1}{6}$

$p = 1.045^{\frac{1}{6}} - 1 = 1.0073631 - 1 = 0.73631\%$

$$34\,200.11 = \text{PMT}\left[\frac{1 - 1.0073631^{-264}}{0.0073631}\right]$$

$$34\,200.11 = 116.23180\text{PMT}$$

$$\text{PMT} = \$294.24$$

3. (a) $\text{PV}_{nc} = 10\,000.00$; $\text{PMT} = 950.00$; $i = 2.5\%$; $c = \dfrac{4}{2} = 2$

$p = 1.025^2 - 1 = 1.050625 - 1 = 5.0625\%$

$$10\,000.00 = 950.00\left[\frac{1 - 1.050625^{-n}}{0.050625}\right]$$

$$10.526316 = \frac{1 - 1.050625^{-n}}{0.050625}$$

$$0.5328947 = 1 - 1.050625^{-n}$$

$$1.050625^{-n} = 0.4671053$$

$$-n\ln 1.050625 = \ln 0.4671053$$

$$-0.049385225n = -0.7612006$$

$$n = 15.413529$$

The number of payments is 16.

(b) $\text{PMT} = 950.00$; $p = 5.0625\%$; $n = 15.413529 - 12 = 3.413529$

$$\text{PV}_{12} = 950.00\left[\frac{1 - 1.050625^{-3.413529}}{0.050625}\right]$$

$$= 950.00(3.0643862)$$

$$= \$2911.17$$

(c) Total paid $= 950.00(12) = \$11\,400.00$
Principal repaid $= 10\,000.00 - 2911.17 = \$7088.83$
Cost of loan $= 11\,400.00 - 7088.83 = \$4311.17$

5. $PV_{nc} = 16\,000.00$; $i = 2.25\%$; $n = 7$; $c = 4$

$p = 1.0225^4 - 1 = 1.0930833 - 1 = 9.30833\%$

$$16\,000.00 = PMT\left[\frac{1 - 1.0930833^{-7}}{0.0930833}\right]$$

$16\,000.00 = 4.9813023\,PMT$

   $PMT = \$3212.01$

Amortization Schedule

Payment number	Amount paid	Interest paid	Principal repaid	Outstanding principal
0				16 000.00
1	3 212.01	1 489.33	1 722.68	14 277.32
2	3 212.01	1 328.98	1 883.03	12 394.29
3	3 212.01	1 153.70	2 058.31	10 335.98
4	3 212.01	962.11	2 249.90	8 086.08
5	3 212.01	752.68	2 459.33	5 626.75
6	3 212.01	523.76	2 688.25	2 938.50
7	3 212.03	273.53	2 938.50	—
Total	22 484.09	6 484.09	16 000.00	

7. $PMT = 3212.01$; $p = 9.30833\%$; $n = 7 - 4 = 3$

$$PV_4 = 3212.01\left[\frac{1 - 1.0930833^{-3}}{0.0930833}\right]$$

$$= 3212.01(2.5174493)$$

$$= \$8086.07$$

Interest paid in $PMT_5 = 8086.07(0.0930833) = \$752.68$

9.   (a) $PV_{nc} = 40\,000.00$; $i = 4.25\%$; $n = 300$; $c = \dfrac{2}{12} = \dfrac{1}{6}$

       $p = 1.0425^{\frac{1}{6}} - 1 = 1.00696106 - 1 = 0.696106\%$

$$40\,000.00 = PMT\left[\frac{1 - 1.00696106^{-300}}{0.00696106}\right]$$

       $40\,000.00 = 125.728700\,PMT$

          $PMT = \$318.15$

   (b) $PMT = 318.15$; $p = 0.696106\%$; $n = 300 - 12 = 288$

$$PV_{12} = 318.15\left[\frac{1 - 1.00696106^{-288}}{0.00696106}\right]$$

$$= 318.15(124.172475)$$

$$= \$39\,505.47$$

       Total paid $= 318.15(12) = \$3817.80$

       Principal repaid $= 40\,000.00 - 39\,505.47 = \$494.53$

       Interest paid $= 3817.80 - 494.53 = \$3323.27$

(c)   PMT $= 318.15$; $p = 0.696106\%$; $n = 300 - 47 = 253$

$$PV_{47} = 318.15\left[\frac{1 - 1.00696106^{-253}}{0.00696106}\right]$$

$$= 318.15(118.81823)$$

$$= \$37\,801.89$$

Interest paid in $PMT_{48} = 37\,801.89(0.0069611) = \$263.14$

(d)   PMT $= 318.15$; $p = 0.696106\%$; $n = 300 - 60 = 240$

$$PV_{60} = 318.15\left[\frac{1 - 1.00696106^{-240}}{0.00696106}\right]$$

$$= 318.15(116.47423)$$

$$= \$37\,056.28$$

$$PV_{nc} = 37\,056.28; \ i = 5.25\%; \ n = 240; \ c = \frac{1}{6}$$

$$p = 1.0525^{\frac{1}{6}} - 1 = 1.0085645 - 1 = 0.85645\%$$

$$37\,056.28 = PMT\left[\frac{1 - 1.0085645^{-240}}{0.0085645}\right]$$

$$37\,056.28 = 101.680561PMT$$

$$PMT = \$364.44$$

(e)

Partial Amortization Schedule

Payment number	Amount paid	Interest paid	Principal repaid	Outstanding principal
0				40 000.00
1	318.15	278.44	39.71	39 960.29
2	318.15	278.17	39.98	39 920.31
3	318.15	277.89	40.26	39 880.05
⋮	⋮	⋮	⋮	⋮
⋮	⋮	⋮	⋮	⋮
60	⋮	⋮	⋮	37 056.28
61	364.44	317.37	47.07	37 009.21
62	364.44	316.97	47.47	36 961.74
63	364.44	316.56	47.88	36 913.86
⋮	⋮	⋮	⋮	⋮

Exercise **14.3**

A. 1. $PV_n = 17\,500.00;\ PMT = 1100.00;\ i = 2.25\%$

$$17\,500.00 = 1100.00\left[\frac{1 - 1.0225^{-n}}{0.0225}\right]$$

$$15.909091 = \frac{1 - 1.0225^{-n}}{0.0225}$$

$$0.3579545 = 1 - 1.0225^{-n}$$

$$1.0225^{-n} = 0.6420455$$

$$-n\ln 1.0225 = \ln 0.6420455$$

$$-0.0222506n = -0.4430961$$

$$n = 19.913895$$

By Method 1 (direct):

$PMT = 1100.00;\ i = 2.25\%;\ n = 0.913895$

$$PV_n = 1100.00\left[\frac{1 - 1.0225^{-0.913895}}{0.0225}\right]$$

$$= 1100.00(0.8946384)$$

$$= \$984.10$$

Final payment $= 984.10(1.0225) = \$1006.24$

3. $PV_n = 7800.00;\ PMT = 775.00(\text{due});\ i = 3.5\%$

$$7800.00 = 775\left[\frac{1 - 1.035^{-n}}{0.035}\right](1.035)$$

$$9.7241701 = \frac{1 - 1.035^{-n}}{0.035}$$

$$1.035^{-n} = 0.6596540$$

$$-n\ln 1.035 = \ln 0.6596540$$

$$-0.0344014n = -0.4160398$$

$$n = 12.0936787$$

Present value of final payment

$PMT = 775.00;\ i = 3.5\%;\ n = 0.0936787$

$$PV_n(\text{due}) = 775.00(1.035)\left[\frac{1 - 1.035^{-0.0936787}}{0.035}\right]$$

$$= 775.00(1.035)(0.0919284)$$

$$= 775.00(0.0951459)$$

$$= \$73.74$$

The final payment is $73.74.

5. $PV_{nc} = 15\,400.00;\ PMT = 1600.00;\ i = 2\%;\ c = \dfrac{4}{2} = 2$

$p = 1.02^2 - 1 = 1.0404 - 1 = 4.04\%$

$$15\,400.00 = 1600.00\left[\frac{1 - 1.0404^{-n}}{0.0404}\right]$$

$$9.625 = \frac{1 - 1.0404^{-n}}{0.0404}$$

$$1.0404^{-n} = 0.61115$$
$$-n\ln 1.0404 = \ln 0.61115$$
$$-0.0396053n = -0.4924129$$
$$n = 12.433019$$

Present value of final payment:

$PMT = 1600.00;\ p = 4.04\%;\ n = 0.433019$

$$PV_n = 1600.00\left[\frac{1 - 1.0404^{-0.433019}}{0.0404}\right]$$

$$= 1600.00(0.4208814)$$

$$= \$673.41$$

Final payment $= 673.41(1.0404) = \$700.62$

7. $PV_{nc} = 17\,300.00;\ PMT = 425.00(\text{due});\ i = 1.5\%;\ c = \dfrac{4}{12} = \dfrac{1}{3}$

$p = 1.015^{\frac{1}{3}} - 1 = 1.0049752 - 1 = 0.49752\%$

$$17\,300.00 = 425.00(1.0049752)\left[\frac{1 - 1.0049752^{-n}}{0.0049752}\right]$$

$$40.504365 = \frac{1 - 1.0049752^{-n}}{0.0049752}$$

$$1.0049752^{-n} = 0.7984827$$
$$-n\ln 1.0049752 = \ln 0.7984827$$
$$-0.0049629n = -0.22504198$$
$$n = 45.344859$$

Present value of final payment:

$PMT = 425.00;\ p = 0.49752\%;\ n = 0.344859$

$$PV_{nc}(\text{due}) = 425.00(1.0049752)\left[\frac{1 - 1.0049752^{-0.344859}}{0.0049752}\right]$$

$$= 425.00(1.0049752)(0.3437098)$$

$$= 425.00(0.3454192)$$

$$= \$146.80$$

B.   1.   (a) $PV_n = 7200.00$; $PMT = 360.00$; $i = 2.75\%$

$$7200.00 = 360.00\left[\frac{1 - 1.0275^{-n}}{0.0275}\right]$$

$$20.00 = \frac{1 - 1.0275^{-n}}{0.0275}$$

$$1.0275^{-n} = 0.45$$
$$-n \ln 1.0275 = \ln 0.45$$
$$-0.0271287n = -0.7985077$$
$$n = 29.434057$$

Number of payments is 30.

(b)   Present value of final payment:
$PMT = 360.00$; $i = 2.75\%$; $n = 0.434057$

$$PV_n = 360.00\left[\frac{1 - 1.0275^{-0.434057}}{0.0275}\right]$$

$$= 360.00(0.4256847)$$
$$= \$153.25$$

Final payment $= 153.25(1.0275) = \$157.46$

3. $PV_{nc} = 25\,000.00$; $PMT = 1200.00$; $i = 0.5\%$; $c = \dfrac{12}{4} = 3$

$p = 1.005^3 - 1 = 1.0150751 - 1 = 1.50751\%$

$$25\,000.00 = 1200.00\left[\frac{1 - 1.0150751^{-n}}{0.0150751}\right]$$

$$20.833333 = \frac{1 - 1.0150751^{-n}}{0.0150751}$$

$$1.0150751^{-n} = 0.6859354$$
$$-n \ln 1.0150751 = \ln 0.6859354$$
$$-0.0149626n = -0.3769718$$
$$n = 25.194271$$

Present value of final payment:
$PMT = 1200.00$; $p = 1.50751\%$; $n = 0.194271$

$$PV_{nc} = 1200.00\left[\frac{1 - 1.0150751^{-0.194271}}{0.0150751}\right]$$

$$= 1200.00(0.192541)$$
$$= \$231.05$$

Final payment $= 231.05(1.0150751) = \$234.53$

5. $PV_{nc}(\text{due}) = 20\,000.00; \ PMT = 1000.00; \ i = 1.75\%$

$$20\,000.00 = 1000.00(1.0175)\left[\frac{1 - 1.0175^{-n}}{0.0175}\right]$$

$$19.656020 = \frac{1 - 1.0175^{-n}}{0.0175}$$

$$1.0175^{-n} = 0.6560197$$

$$-n\ln 1.0175 = \ln 0.6560197$$

$$-0.0173486n = -0.4215645$$

$$n = 24.299626$$

Present value of final payment:

$PMT = 1000.00; \ i = 1.75\%; \ n = 0.299626$

$$PV_{nc}(\text{due}) = 1000.00(1.0175)\left[\frac{1 - 1.0175^{-0.299626}}{0.0175}\right]$$

$$= 1000.00(1.0175)(0.2962638)$$

$$= 1000.00(0.3014484)$$

$$= \$301.45$$

The final payment is $301.45.

7. $PV_{nc}(\text{due}) = 42\,000.00; \ PMT = 5000.00; \ i = 2.25\%; \ c = \frac{4}{2} = 2$

$$p = 1.0225^2 - 1 = 1.04550625 - 1 = 4.550625\%$$

$$42\,000.00 = 5000.00(1.04550625)\left[\frac{1 - 1.04550625^{-n}}{0.04550625}\right]$$

$$8.0343853 = \frac{1 - 1.04554625^{-n}}{0.04550625}$$

$$1.04550625^{-n} = 0.6343853$$

$$-n\ln 1.04550625 = \ln 0.6343853$$

$$-0.04450122n = -0.45509878$$

$$n = 10.226659$$

Present value of final payment:

$PMT = 5000.00; \ p = 4.550625\%; \ n = 0.226659$

$$PV_{nc}(\text{due}) = 5000.00(1.04550625)\left[\frac{1 - 1.04550625^{-0.226659}}{0.04550625}\right]$$

$$= 5000.00(1.04550625)(0.2205390)$$

$$= 5000.00(0.2305749)$$

$$= \$1152.87$$

The final payment is $1152.87.

9. (a) $PV_n(\text{defer}) = 16\,000.00$; $PMT = 1375.00$; $i = 2.5\%$; $d = 40$

$$16\,000.00 = 1375.00\left(1.025^{-40}\right)\left[\frac{1 - 1.025^{-n}}{0.025}\right]$$

$$16\,000.00 = 1375.00(0.3724306)\left[\frac{1 - 1.025^{-n}}{0.025}\right]$$

$$31.244381 = \frac{1 - 1.025^{-n}}{0.025}$$

$$1.025^{-n} = 0.2188905$$

$$-n\ln 1.025 = \ln 0.2188905$$

$$-0.024692613n = -1.51918367$$

$$n = 61.523813$$

The number of payments is 62.

(b) Present value of final payment:
$PMT = 1375.00$; $i = 2.5\%$; $n = 0.523813$

$$PV_n = 1375.00\left[\frac{1 - 1.025^{-0.523813}}{0.025}\right]$$

$$= 1375.00(0.5140409)$$

$$= \$706.81$$

Final payment $= 706.81(1.025) = \$724.48$

(c) Amount received $= 1375.00(61) + 724.48$
$$= 83\,875.00 + 724.48$$
$$= \$84\,599.48$$

(d) Interest $= 84\,599.48 - 16\,000.00 = \$68\,599.48$

Exercise 14.4

A. 1. (a) $PV_{nc} = 90\,000.00$; $n = 12(25) = 300$; $i = \dfrac{8.50\%}{2} = 4.25\%$; $c = \dfrac{1}{6}$

$$p = 1.0425^{\frac{1}{6}} - 1 = 1.0069611 - 1 = 0.0069611 = 0.69611\%$$

$$90\,000.00 = PMT\left[\frac{1 - 1.0069611^{-300}}{0.0069611}\right]$$

$$90\,000.00 = PMT(125.72819)$$

$$PMT = 715.83$$

The monthly payment is $715.83.

(b) $PMT = 715.83$; $n = 300 - 60 = 240$; $p = 0.69611\%$

$$PV_{nc} = 715.83\left[\frac{1 - 1.0069611^{-240}}{0.0069611}\right]$$

$$= 715.83(116.47382) = 83\,375.45$$

The balance at the end of the five-year term is \$83 375.45.

(c) $PV_{nc} = 83\,375.45$; $n = 240$; $i = \dfrac{7\%}{2} = 3.5\%$; $c = \dfrac{1}{6}$

$p = 1.035^{\frac{1}{6}} - 1 = 1.005750 - 1 = 0.00575 = 0.575\%$

$$83\,375.45 = PMT\left[\frac{1 - 1.00575^{-240}}{0.00575}\right]$$

$$83\,375.45 = PMT(129.987)$$

$$PMT = 641.41$$

The monthly payment for the renewal term is \$641.41.

3. (a) $PV_{nc} = 40\,000.00$; $n = 12(10) = 120$; $i = \dfrac{9\%}{2} = 4.5\%$; $c = \dfrac{1}{6}$

$p = 1.045^{\frac{1}{6}} - 1 = 1.0073631 - 1 = 0.0073631 = 0.73631\%$

$$40\,000.00 = PMT\left[\frac{1 - 1.0073631^{-120}}{0.0073631}\right]$$

$$40\,000.00 = PMT(79.498581)$$

$$PMT = 503.15$$

The rounded monthly payment is \$550.00.
$PV_{nc} = 40\,000.00$; $PMT = 550.00$; $p = 0.73631\%$

$$40\,000.00 = 550.00\left[\frac{1 - 1.0073631^{-n}}{0.0073631}\right]$$

$$0.5354982 = 1 - 1.0073631^{-n}$$

$$1.0073631^{-n} = 0.4645018$$

$$-n\ln 1.0073631 = \ln 0.4645018$$

$$-0.0073361n = -0.7667898$$

$$n = 104.522813 \text{ (months)}$$

104 rounded payments of \$550.00 plus a final payment less than \$550.00 is required to repay the mortgage.

(b) $PMT = 550.00$; $n = 104.522813 - 104 = 0.522813$; $p = 0.73631\%$

$$PV_{nc} = 550.00\left[\frac{1 - 1.0073631^{-0.522813}}{0.0073631}\right]$$

$$= 550.00(0.51990) = 285.94$$

The size of the last payment $= 285.94(1.0073631) = \$288.05$

(c) The total amount paid with unrounded payments
= 120(503.15) = $60 378.00
The total amount paid with rounded payments
= 104(550.00) + 288.05 = 57 200.00 + 288.05 = $57 488.05
Amount of interest saved = 60 378.00 − 57 488.05 = $2889.95

5. $PV_{nc} = 80\,000.00$; PMT $= 826.58$; $n = 12(15) = 180$; $c = \dfrac{1}{6}$; $m = 2$

0	80 000	826.58	180			
FV	PV	PMT	N	CPT	%i	0.7764367

$p = 0.7764367\%$
The effective monthly rate is 0.7764367%.

$$1.0077644 = (1 + i)^{\frac{1}{6}}$$
$$1.0077644^6 = 1 + i$$
$$1 + i = 1.047500$$
$$i = 0.0475 = 4.75\%$$

The semi-annual rate is 4.75%.
The nominal annual rate compounded semi-annually $= 2(4.75\%) = 9.50\%$

7. (a) $PV_{nc} = 105\,000.00$; $n = 12(20) = 240$; $i = \dfrac{7.25\%}{2} = 3.625\%$; $c = \dfrac{1}{6}$

$p = 1.03625^{\frac{1}{6}} - 1 = 1.0059524 - 1 = 0.0059524 = 0.59524\%$

$$105\,000.00 = PMT\left[\frac{1 - 1.0059524^{-240}}{0.0059524}\right]$$

$105\,000.00 = PMT(127.56754)$

$\quad\quad PMT = 823.09$

The monthly payment is $823.09.
PMT $= 823.09$; $n = 240 - 36 = 204$; $p = 0.59524\%$

$$PV_{nc} = 823.09\left[\frac{1 - 1.0059524^{-204}}{0.0059524}\right]$$

$\quad\quad = 823.09(117.93703) = 97\,072.79$

The balance at the end of the third year
= 97 072.79 − 7000.00 = $90 072.79

To determine the balance at the end of the four-year term, use the Retrospective Method:
The accumulated value of $90 072.79 at the end of four years

$= 90\,072.79\left(1.0059524^{12}\right) = 90\,072.79(1.0738143) = \$96\,721.45$

The accumulated value of the payments made during the fourth year

$$FV_{nc} = 823.09\left[\frac{1.0059524^{12} - 1}{0.0059524}\right]$$

$\quad\quad = 823.09(12.400759) = 10\,206.94$

The mortgage balance at the end of the four-year term
= 96 721.45 − 10 206.94 = $86 514.51

(b) $PV_{nc} = 86\,514.51$; $PMT = 823.09$; $p = 0.59524\%$

$$86\,514.51 = 823.09\left[\frac{1 - 1.0059524^{-n}}{0.0059524}\right]$$

$$0.6256533 = 1 - 1.0059524^{-n}$$

$$1.0059524^{-n} = 0.3743467$$

$$-n\ln 1.0059524 = \ln 0.3743467$$

$$-0.0059348n = -0.9825729$$

$$n = 165.56125 \text{ (months)}$$

165 payments of \$823.09 plus a final payment less than \$823.09 are required to repay the mortgage.

(c) $PMT = 823.09$; $n = 165.56125 - 165 = 0.56125$; $p = 0.59524\%$

$$PV_{nc} = 823.09\left[\frac{1 - 1.0059524^{-0.56125}}{0.0059524}\right]$$

$$= 823.09(0.5586553) = 459.82$$

The size of the last payment $= 459.82(1.0059524) = \$462.56$
The total amount paid with contractual payments $= 240(823.09) = \$197\,541.60$
The total amount paid with the lump-sum payment
$= 165(823.09) + 462.56 + 7000.00 = 135\,809.85 + 7462.56 = \$143\,272.41$
Difference in cost $= 197\,541.60 - 143\,272.41 = \$54\,269.19$

9. $PV_{nc} = 40\,000.00$; $n = 12(12) = 144$; $i = \dfrac{5.5\%}{2} = 2.75\%$; $c = \dfrac{1}{6}$

$p = (1.0275)^{\frac{1}{6}} - 1 = 1.0045317 - 1 = 0.0045317 = 0.45317\%$

$$40\,000.00 = PMT\left[\frac{1 - 1.0045317^{-144}}{0.0045317}\right]$$

$$40\,000.00 = PMT(105.59465)$$

$$PMT = 378.81$$

The rounded monthly payment is \$380.00.

Amortization Schedule for the First 6 Months

Payment date	Amount paid	Interest paid	Principal repaid	Balance
June 1				40 000.00
July 1	380.00	181.27	198.73	39 801.27
Aug 1	380.00	180.37	199.63	39 601.64
Sept 1	380.00	179.46	200.54	39 401.10
Oct 1	380.00	178.55	201.45	39 199.65
Nov 1	380.00	177.64	202.36	38 997.29
Dec 1	380.00	176.72	203.28	38 794.01

11. Mortgage statement:

PMT = 190.00; annual rate of interest = 12(0.45317%) = 5.43804%

Payment date	Number of days	Amount paid	Interest paid	Principal	Balance repaid
June 1					40 000.00
16	15	190.00	89.39	100.61	39 899.39
July 1	15	190.00	89.17	100.83	39 798.56
16	15	190.00	88.94	101.06	39 697.50
Aug 1	16	190.00	94.63	95.37	39 602.13
16	15	190.00	88.50	101.50	39 500.63
Sept 1	16	190.00	94.16	95.84	39 404.79
16	15	190.00	88.06	101.94	39 302.85
Oct 1	15	190.00	87.83	102.17	39 200.68
15	15	190.00	87.61	102.39	39 098.29
Nov 1	16	190.00	93.20	96.80	39 001.49
15	15	190.00	87.16	102.84	38 898.65
Dec 1	15	190.00	86.93	103.07	38 795.58

The mortgage statement balance on December 1 of $38 795.58 differs from the amortization schedule balance of $38 794.01 by $1.57. The difference is reduced from $3.01 in the answer to Question 10 by $1.44 due to making semi-monthly payments.

Review Exercise

1. (a) $PV_n = 45\,000.00 - 10\,000.00 = 35\,000.00$; $i = 2\%$; $n = 32$

$$35\,000.00 = PMT\left[\frac{1 - 1.02^{-32}}{0.02}\right]$$

$$35\,000.00 = 23.468335\,PMT$$

$$PMT = \$1491.37$$

(b) Total paid = 1491.37(32)   = $47 723.84

Amount borrowed   = 35 000.00

   Cost of financing   = $12 723.84

(c) Balance after 5 years: $n = 32 - 20 = 12$

$$PV_{20} = 1491.37\left[\frac{1 - 1.02^{-12}}{0.02}\right]$$

$$= 1491.37(10.575341)$$

$$= \$15\,771.75$$

(d) Balance after 19 payments: $n = 32 - 19 = 13$

$$PV_{19} = 1491.37\left[\frac{1 - 1.02^{-13}}{0.02}\right]$$

$$= 1491.37(11.348374)$$
$$= \$16\,924.62$$

Interest in 20th payment $= 16\,924.62(0.02) = \$338.49$

(e) Balance after 23 payments: $n = 32 - 23 = 9$

$$PV_{23} = 1491.37\left[\frac{1 - 1.02^{-9}}{0.02}\right]$$

$$= 1491.37(8.1622367)$$
$$= \$12\,172.91$$

Interest in 24th payment $= 12\,172.91(0.02) = \$243.46$
Principal repaid $= 1491.37 - 243.46 = \$1247.91$

(f) Balance after 9 payments: $n = 32 - 9 = 23$

$$PV_9 = 1491.37\left[\frac{1 - 1.02^{-23}}{0.02}\right] = 1491.37(18.292204) = \$27\,280.44$$

Balance after 29 payments: $n = 32 - 29 = 3$

$$PV_{29} = 1491.37\left[\frac{1 - 1.02^{-3}}{0.02}\right] = 1491.37(2.8838833) = \$4300.94$$

Partial Amortization Schedule

Payment number	Periodic payment	Interest paid	Principal repaid	Outstanding balance
0				35 000.00
1	1491.37	700.00	791.37	34 208.63
2	1491.37	684.17	807.20	33 401.43
3	1491.37	668.03	823.34	32 578.09
⋮	⋮	⋮	⋮	⋮
9	1491.37	⋮	⋮	27 280.44
10	1491.37	545.61	945.76	26 334.68
11	1491.37	526.69	964.68	25 370.00
12	1491.37	507.40	983.97	24 386.03
⋮	⋮	⋮	⋮	⋮
29	1491.37	⋮	⋮	4 300.94
30	1491.37	86.02	1 405.35	2 895.59
31	1491.37	57.91	1 433.46	1 462.13
32	1491.37	29.24	1 462.13	—
Total	47 723.84	12 723.84	35 000.00	

3. (a) $PV_n = 40\,000.00$; $PMT = 2000.00$; $i = 1.75\%$

$$40\,000.00 = 2000.00\left[\frac{1 - 1.0175^{-n}}{0.0175}\right]$$

$$20.00 = \frac{1 - 1.0175^{-n}}{0.0175}$$

$$1.0175^{-n} = 0.65$$
$$-n\ln 1.0175 = \ln 0.65$$
$$-0.0173486n = -0.4307829$$
$$n = 24.830989$$

The number of payments is 25.

(b) Balance after 2 years: $n = 24.830989 - 8 = 16.830989$

$$PV_8 = 2000.00\left[\frac{1 - 1.0175^{-16.830989}}{0.0175}\right]$$
$$= 2000.00(14.470145)$$
$$= \$28\,940.29$$

(c) Balance after 11th payment: $n = 24.830989 - 11 = 13.830989$

$$PV_{11} = 2000.00\left[\frac{1 - 1.0175^{-13.830989}}{0.0175}\right]$$
$$= 2000.00(12.190393)$$
$$= \$24\,380.79$$

Interest in 12th payment $= 24\,380.79(0.0175) = \$426.66$

(d) Balance after 19th payment: $n = 24.830989 - 19 = 5.830989$

$$PV_{19} = 2000.00\left[\frac{1 - 1.0175^{-5.830989}}{0.0175}\right]$$
$$= 2000.00(5.4977904)$$
$$= \$10\,995.58$$

Interest in 20th payment $= 10\,995.58(0.0175) = \$192.42$
Principal repaid $= 2000.00 - 192.42 = \$1807.58$

(e)   The last 3 payments are PMT_{23}, PMT_{24}, PMT_{25}.

Balance after 22nd payment: $n = 24.830989 - 22 = 2.830989$

$$PV_{22} = 2000.00\left[\frac{1 - 1.0175^{-2.830989}}{0.0175}\right]$$

$$= 2000.00(2.7386987)$$

$$= \$5477.40$$

Partial Amortization Schedule

Payment number	Periodic payment	Interest paid	Principal repaid	Outstanding balance
0				40 000.00
1	2 000.00	700.00	1 300.00	38 700.00
2	2 000.00	677.25	1 322.75	37 377.25
3	2 000.00	654.10	1 345.90	36 031.35
⋮	⋮	⋮	⋮	⋮
22	⋮	⋮	⋮	5 477.40
23	2 000.00	95.85	1 904.15	3 573.25
24	2 000.00	62.53	1 937.47	1 635.78
25	1 664.41	28.63	1 635.78	—
Total	49 664.41	9664.41	40 000.00	

5.   (a) $PV_{nc} = 27\,500.00$; $i = 7\%$; $n = 60$; $c = \dfrac{1}{4}$

$$p = 1.07^{\frac{1}{4}} - 1 = 1.0170585 - 1 = 1.70585\%$$

$$27\,500.00 = PMT\left[\frac{1 - 1.0170585^{-60}}{0.0170585}\right]$$

$$27\,500.00 = 37.374532PMT$$

$$PMT = \$735.80$$

(b)   Balance after 3 payments: $n = 60 - 3 = 57$

$$PV_3 = 735.80\left[\frac{1 - 1.0170585^{-57}}{0.0170585}\right]$$

$$= 735.80(36.268538)$$

$$= \$26\,686.39$$

Interest in $PMT_4 = 26\,686.39(0.0170585) = \$455.23$

Principal repaid $= 735.80 - 455.23 = \$280.57$

(c)   Balance after 3 years: $60 - 12 = 48$

$$PV_{12} = 735.80\left[\frac{1 - 1.0170585^{-48}}{0.0170585}\right]$$

$$= 735.80(32.5929898)$$

$$= \$23\,981.92$$

(d) $\quad PV_{nc} = 23\,981.92; \quad i = 3.75\%; \quad n = 32; \quad c = \dfrac{2}{4} = \dfrac{1}{2}$

$p = 1.0375^{\frac{1}{2}} - 1 = 1.0185774 - 1 = 1.85774\%$

$23\,981.92 = PMT\left[\dfrac{1 - 1.0185774^{-32}}{0.0185774}\right]$

$23\,981.92 = 23.960861 PMT$

$\quad PMT = \$1000.88$

(e) Balance after 13th payment of 4-year term:
$n = 32 - 13 = 19$

$PV_{19} = 1000.88\left[\dfrac{1 - 1.0185774^{-19}}{0.0185774}\right]$

$\quad = 1000.88(15.886132)$

$\quad = \$15\,900.11$

Total paid after 7 years $= 735.80(12) + 1000.88(16)$

$\quad = 8829.60 + 16\,014.08$

$\quad = \$24\,843.68$

Principal repaid $= 27\,500.00 - 13\,744.06 = \$13\,755.94$

Partial Amortization Schedule

Payment interval	Periodic payment	Interest paid	Principal repaid	Outstanding balance
0				27 500.00
1	735.80	469.11	266.69	27 233.31
2	735.80	464.56	271.24	26 962.07
3	735.80	459.93	275.87	26 686.20
⋮	⋮	⋮	⋮	⋮
12	⋮	⋮	⋮	23 981.92
13	1 000.88	445.52	555.36	23 426.56
14	1 000.88	435.20	565.68	22 860.88
15	1 000.88	424.70	576.18	22 284.70
⋮	⋮	⋮	⋮	⋮
25	⋮	⋮	⋮	15 900.11
26	1 000.88	295.38	705.50	15 194.61
27	1 000.88	282.28	718.60	14 476.01
28	1 000.88	268.93	731.95	13 744.06
Total	24 843.68	11 087.74	13 755.94	

7. (a) $PV_n = 25\,000.00$; $PMT = 3500.00$; $i = 5.5\%$

$$25\,000.00 = 3500.00\left[\frac{1 - 1.055^{-n}}{0.055}\right]$$

$$7.1428571 = \frac{1 - 1.055^{-n}}{0.055}$$

$$1.055^{-n} = 0.60714286$$

$$-n\ln 1.055 = \ln 0.60714286$$

$$-0.053540767n = -0.4989912$$

$$n = 9.3198366$$

The number of payments is 10.

(b) Balance after 9th payment: $n = 9.3198366 - 9 = 0.3198366$

$$PV_9 = 3500.00\left[\frac{1 - 1.055^{-0.3198366}}{0.055}\right]$$

$$= 3500.00(0.3087002)$$

$$= \$1080.45$$

Final payment $= 1080.45(1.055) = \$1139.88$

9. (a) $PV_n(\text{defer}) = 33\,000.00$; $PMT = 4300.00(\text{due})$; $i = 2.5\%$; $d = 12 - 1 = 11$

$$33\,000.00 = 4300.00\left(1.025^{-11}\right)\left[\frac{1 - 1.025^{-n}}{0.025}\right]$$

$$33\,000.00 = 4300.00(0.7621448)\left[\frac{1 - 1.025^{-n}}{0.025}\right]$$

$$10.069502 = \frac{1 - 1.025^{-n}}{0.025}$$

$$1.025^{-n} = 0.7482624$$

$$-n\ln 1.025 = \ln 0.7482624$$

$$-0.024692613n = -0.2900015$$

$$n = 11.744464$$

The number of payments is 12.

(b) Balance after 11 payments: $n = 11.744464 - 11 = 0.744464$

$$PV_{11}(\text{due}) = 4300.00(1.025)\left[\frac{1 - 1.025^{-0.744464}}{0.025}\right]$$

$$= 4300.00(1.025)(0.7285932)$$

$$= \$3211.27$$

11. (a) $\text{PV}_{nc} = 135\,000.00; \; n = 12(25) = 300; \; i = \dfrac{8.70\%}{2} = 4.35\%; \; c = \dfrac{1}{6}$

$p = 1.0435^{\frac{1}{6}} - 1 = 1.0071220 - 1 = 0.71220\%$

$135\,000.00 = \text{PMT}\left[\dfrac{1 - 1.0071220^{-300}}{0.0071220}\right]$

$135\,000.00 = \text{PMT}\,(123.70782)$

$\text{PMT} = 1091.28$

The monthly payment is \$1091.28.

(b) $\text{PMT} = 1091.28; \; n = 300 - 60 = 240; \; p = 0.71220\%$

$\text{PV}_{nc} = 1091.28\left[\dfrac{1 - 1.0071220^{-240}}{0.0071220}\right]$

$\phantom{\text{PV}_{nc}} = 1091.28(114.84192) = 125\,324.69$

The balance at the end of the five-year term is \$125 324.69.

(c) $\text{PV}_{nc} = 125\,324.69; \; n = 240; \; i = \dfrac{7.8\%}{2} = 3.9\%; \; c = \dfrac{1}{6}$

$p = 1.039^{\frac{1}{6}} - 1 = 1.0063968 - 1 = 0.63968\%$

$125\,324.69 = \text{PMT}\left[\dfrac{1 - 1.0063968^{-240}}{0.0063968}\right]$

$125\,324.69 = \text{PMT}(122.48913)$

$\text{PMT} = 1023.15$

The monthly payment for the renewal term is \$1023.15.

13. (a) $\text{PV}_{nc} = 180\,000.00; \; n = 12(10) = 120; \; i = \dfrac{8.5\%}{2} = 4.25\%; \; c = \dfrac{1}{6}$

$p = 1.0425^{\frac{1}{6}} - 1 = 1.0069611 - 1 = 0.69611\%$

$180\,000.00 = \text{PMT}\left[\dfrac{1 - 1.0069611^{-120}}{0.0069611}\right]$

$180\,000.00 = \text{PMT}(81.167130)$

$\text{PMT} = 2217.65$

The rounded monthly payment is \$2250.00.

$\text{PV}_{nc} = 180\,000.00; \; \text{PMT} = 2250.00; \; p = 0.69611\%$

$180\,000.00 = 2250.00\left[\dfrac{1 - 1.0069611^{-n}}{0.0069611}\right]$

$0.5568880 = 1 - 1.0069611^{-n}$

$1.0069611^{-n} = 0.4431120$

$-n \ln 1.0069611 = \ln 0.4431120$

$-0.0069370n = -0.8139327$

$n = 117.33209 \; (\text{months})$

117 rounded payments of \$2250.00 plus a final payment less than \$2250.00 is required to repay the mortgage.

(b) PMT $= 2250.00;\ n = 117.33209 - 117 = 0.33209;\ p = 0.69611\%$

$$PV_{nc} = 2250.00\left[\frac{1 - 1.0069611^{-0.33209}}{0.0069611}\right]$$

$$= 2250.00(0.3305586) = 743.76$$

The size of the last payment $= 743.76(1.0069611) = \$748.93$

(c) The total amount paid with unrounded payments
$= 120(2217.65) = \$266\ 118.00$
The total amount paid with rounded payments
$= 117(2250.00) + 748.93 = 263\ 250.00 + 748.93 = \$263\ 998.93$
Amount of interest saved $= 266\ 118.00 - 263\ 998.93 = \$2119.07$

15. (a) $PV_n = 6500.00;\ i = 0.75\%;\ n = 48$

$$6500.00 = PMT\left[\frac{1 - 1.0075^{-48}}{0.0075}\right]$$

$$6500.00 = 40.184782PMT$$

$$PMT = \$161.75$$

(b) Total paid $= 161.75(48) = \$7764.00$
Original principal $\quad= \underline{\ \ 6500.00}$
Total interest $\quad\quad= \$1264.00$

(c) Balance after 1 year: $n = 48 - 12 = 36$

$$PV_{12} = 161.75\left[\frac{1 - 1.0075^{-36}}{0.0075}\right]$$

$$= 161.75(31.446805)$$

$$= \$5086.52$$

(d) Balance after 29th payment: $n = 48 - 29 = 19$

$$PV_{29} = 161.75\left[\frac{1 - 1.0075^{-19}}{0.0075}\right]$$

$$= 161.75(17.646830)$$

$$= \$2854.37$$

Interest in $PMT_{30} = 2854.37(0.0075) = \$21.41$

(e)   Last 3 payments are PMT_{46}, PMT_{47} and PMT_{48}.

Balance after 45 payments: $n = 3$

$$PV_{45} = 161.75\left[\frac{1 - 1.0075^{-3}}{0.0075}\right]$$

$$= 161.75(2.9555562)$$

$$= \$478.06$$

Partial Amortization Schedule

Payment interval	Periodic payment	Interest paid	Principal repaid	Outstanding balance
0				6500.00
1	161.75	48.75	113.00	6387.00
2	161.75	47.90	113.85	6273.15
3	161.75	47.05	114.70	6158.45
⋮	⋮	⋮	⋮	⋮
45	⋮	⋮	⋮	478.06
46	161.75	3.59	158.16	319.90
47	161.75	2.40	159.35	160.55
48	161.75	1.20	160.55	—
Total	7764.00	1264.00	6500.00	

17. $PV_{nc} = 95\,000.00$; $\text{PMT} = 748.06$; $n = 12(25) = 300$; $c = \dfrac{1}{6}$; $m = 2$

0	95 000	748.06	300			
FV	PV	PMT	N	CPT	%i	0.6862295

The effective monthly rate is 0.6862295%.

$$1.0068623 = (1 + i)^{\frac{1}{6}}$$

$$1.0068623^6 = 1 + i$$

$$1 + i = 1.04188666$$

$$i = 0.04188666 = 4.189\%$$

The semi-annual rate is 4.189%.

The nominal annual rate compounded semi-annually $= 2(4.189\%) = 8.38\%$.

19.   (a) $PV_{nc} = 28\,000.00$; $i = 3\%$; $n = 80$; $c = \dfrac{2}{4} = \dfrac{1}{2}$

$$p = 1.03^{\frac{1}{2}} - 1 = 1.0148892 - 1 = 1.448892\%$$

$$28\,000.00 = \text{PMT}\left[\frac{1 - 1.0148892^{-80}}{0.0148892}\right]$$

$$28\,000.00 = 46.573638\text{PMT}$$

$$\text{PMT} = \$601.20$$

(b)   Balance after 1 year: $n = 80 - 4 = 76$

$$PV_{76} = 601.20\left[\frac{1 - 1.0148892^{-76}}{0.0148892}\right]$$

$$= 601.20(45.319756)$$

$$= \$27\,246.24$$

Total paid $= 601.20(4)$ $\qquad\qquad = \$2404.80$

Principal repaid $= 28\,000.00 - 27\,246.24 = \underline{\quad 753.76}$

Interest paid $\qquad\qquad\qquad\qquad = \$1651.04$

(c)   Balance after 3 years: $n = 80 - 12 = 68$

$$PV_{68} = 601.20\left[\frac{1 - 1.0148892^{-68}}{0.0148892}\right]$$

$$= 601.20(42.578256)$$

$$= \$25\,598.05$$

(d) $PV_{nc} = 25\,598.05$; $i = 7\%$; $n = 68$; $c = \dfrac{1}{4}$

$$p = 1.07^{\frac{1}{4}} - 1 = 1.0170585 - 1 = 1.70585\%$$

$$25\,598.05 = PMT\left[\frac{1 - 1.0170585^{-68}}{0.0170585}\right]$$

$$25\,598.05 = 40.063609\,PMT$$

$$PMT = 638.94$$

Self-Test

1. $PV_n = 9000.00$; $i = 1\%$; $n = 60$

$$9000.00 = PMT\left[\frac{1 - 1.01^{-60}}{0.01}\right]$$

$$9000.00 = 44.955038\,PMT$$

$$PMT = \$200.20$$

After 2 years: $n = 36$; $PMT = 200.20$; $i = 1\%$

$$PV_n = 200.20\left[\frac{1 - 1.01^{-36}}{0.01}\right] = 200.20(30.107505) = \$6027.52$$

3. $PV_{nc} = 50\,000.00$; $i = 4.5\%$; $n = 240$; $c = \dfrac{2}{12} = \dfrac{1}{6}$

$p = 1.045^{\frac{1}{6}} - 1 = 1.0073631 - 1 = 0.73631\%$

$$50\,000.00 = PMT\left[\dfrac{1 - 1.0073631^{-240}}{0.0073631}\right]$$

$50\,000.00 = 112.46219\,PMT$

$PMT = 444.59$

Balance after 3 years: $PMT = 444.59$; $p = 0.73631\%$; $n = 204$

$$PV_n = 444.59\left[\dfrac{1 - 1.0073631^{-204}}{0.0073631}\right] = 444.59(105.40439) = 46\,861.74$$

Total paid in 3 years $= 444.59 \times 36$ $\qquad = \$16\,005.24$

Principal repaid after 3 years $= 50\,000.00 - 46\,861.74 = \underline{\ \ 3\,138.26}$

Interest paid $\qquad\qquad\qquad\qquad\qquad\qquad = \$12\,866.98$

5. (a) $PV_{nc} = 190\,000.00$; $n = 12(20) = 240$; $i = \dfrac{6.50\%}{2} = 3.25\%$; $c = \dfrac{1}{6}$

$p = 1.0325^{\frac{1}{6}} - 1 = 1.00534474 - 1 = 0.534474\%$

$$190\,000.00 = PMT\left[\dfrac{1 - 1.00534474^{-240}}{0.00534474}\right]$$

$190\,000.00 = PMT(135.043814)$

$PMT = 1406.95$

The monthly payment is \$1406.95.

(b) $PMT = 1406.95$; $n = 240 - 36 = 204$; $p = 0.534474\%$

$$PV_{nc} = 1406.95\left[\dfrac{1 - 1.00534474^{-204}}{0.00534474}\right]$$

$= 1406.95(124.03150) = 174\,506.12$

The balance at the end of the three-year term is \$174\,506.12.

(c) $PV_{nc} = 174\,506.12$; $n = 204$; $i = \dfrac{7.25\%}{2} = 3.625\%$; $c = \dfrac{1}{6}$

$p = 1.03625^{\frac{1}{6}} - 1 = 1.0059524 - 1 = 0.59524\%$

$$174\,506.12 = PMT\left[\dfrac{1 - 1.0059524^{-204}}{0.0059524}\right]$$

$174\,506.12 = PMT(117.93703)$

$PMT = 1479.66$

The monthly payment for the renewal term is \$1479.66.

7. $PV_{nc} = 145\,000.00$; $PMT = 1297.00$; $n = 12(25) = 300$; $c = \dfrac{1}{6}$; $m = 2$

0	145 000	1297.00	300			
FV	PV	PMT	N	CPT	%i	0.8164793

The effective monthly rate is 0.8164793%.

$1.0081648 = (1 + i)^{\frac{1}{6}}$

$1.0081648^{6} = 1 + i$

$\qquad 1 + i = 1.0499997$

$\qquad\quad i = 0.499997 = 5.00\%$

The semi-annual rate is 5.00%.

The nominal annual rate compounded semi-annually $= 2(5.0\%) = 10.0\%$

15 Bond valuation and sinking funds

Exercise 15.1

A. 1. FV = 100 000.00; PMT = 100 000.00(0.035) = 3500.00; $n = 11$; $i = 3.75\%$

$$\text{Purchase price} = 100\,000.00\left(1.0375^{-11}\right) + 3500.00\left[\frac{1 - 1.0375^{-11}}{0.0375}\right]$$

$$= 100\,000.00(0.6670077) + 3500.00(8.8797949)$$

$$= 66\,700.77 + 31\,079.28$$

$$= \$97\,780.05$$

3. FV = 25 000.00(1.03) = 25 750.00; PMT = 25 000.00(0.03) = 750.00; $n = 14$; $i = 3.5\%$

$$\text{Purchase price} = 25\,750.00\left(1.035^{-14}\right) + 750.00\left[\frac{1 - 1.035^{-14}}{0.035}\right]$$

$$= 25\,750.00(0.6177818) + 750.00(10.9205203)$$

$$= 15\,907.88 + 8190.39$$

$$= \$24\,098.27$$

5. FV = 50 000.00; PMT = 50 000.00(0.0325) = 1625.00; $n = 20$; $i = 6\%$; $c = \frac{1}{2}$

$$p = 1.06^{\frac{1}{2}} - 1 = 1.0295630 - 1 = 2.95630\%$$

$$\text{Purchase price} = 50\,000.00\left(1.0295630^{-20}\right) + 1625.00\left[\frac{1 - 1.0295630^{-20}}{0.0295630}\right]$$

$$= 50\,000.00(0.5583949) + 1625.00(14.937762)$$

$$= 27\,919.74 + 24\,273.86$$

$$= \$52\,193.60$$

7. FV = 8000.00(1.04) = 8320.00; PMT = 8000.00(0.04) = 320.00;

$n = 37$; $i = 0.75\%$; $c = \frac{12}{2} = 6$

$$p = 1.0075^{6} - 1 = 1.0458522 - 1 = 4.58522\%$$

$$\text{Purchase price} = 8320.00\left(1.0458522^{-37}\right) + 320.00\left[\frac{1 - 1.0458522^{-37}}{0.0458522}\right]$$

$$= 8320.00(0.1903699) + 320.00(17.657387)$$

$$= 1583.88 + 5650.36$$

$$= \$7234.24$$

9. FV = 15 000.00; PMT = 15 000.00(0.0475) = 712.50; i = 5%
Interest date preceding the purchase date is 6.5 years before maturity: n = 13

$$\text{Purchase price} = 15\,000.00\left(1.05^{-13}\right) + 712.50\left[\frac{1 - 1.05^{-13}}{0.05}\right]$$

$$= 15\,000.00(0.5303214) + 712.50(9.393573)$$

$$= 7954.82 + 6692.92$$

$$= \$14\,647.74$$

The accumulated value 2 months later:

$$PV = 14\,647.74;\ r = 10\%;\ t = \frac{2}{12} = \frac{1}{6}$$

$$FV = 14\,647.74\left(1 + 0.10 \times \frac{1}{6}\right)$$

$$= 14\,647.74(1.0166667)$$

$$= \$14\,891.87$$

11. FV = 10 000.00(1.08) = 10 800.00; PMT = 10 000.00(0.045) = 450.00
The interest date preceding the date of the purchase is 9.5 years before maturity:
n = 19; i = 4.25%

$$\text{Purchase price} = 10\,800.00\left(1.0425^{-19}\right) + 450.00\left[\frac{1 - 1.0425^{-19}}{0.0425}\right]$$

$$= 10\,800.00(0.4534765) + 450.00(12.859376)$$

$$= 4897.55 + 5786.72$$

$$= \$10\,684.27$$

Accumulated value 1 month later:

$$PV = 10\,684.27;\ r = 8.5\%;\ t = \frac{1}{12}$$

$$FV = 10\,684.27\left(1 + 0.085 \times \frac{1}{12}\right)$$

$$= 10\,684.27(1.00708333)$$

$$= \$10\,759.95$$

B. 1. FV = 500.00; PMT = 500.00(0.03) = \$15.00; i = 3.75%
The purchase date is 5.5 years before maturity: n = 11

$$\text{Purchase price} = 500.00\left(1.0375^{-11}\right) + 15.00\left[\frac{1 - 1.0375^{-11}}{0.0375}\right]$$

$$= 500.00(0.6670077) + 15.00(8.8797949)$$

$$= 333.50 + 133.20$$

$$= \$466.70$$

3. $FV = 1000.00(1.04) = 1040.00$; $PMT = 1000.00(0.045) = 45.00$; $i = 3.25\%$; $n = 17$

$$\text{Purchase price} = 1040.00\left(1.0325^{-17}\right) + 45.00\left[\frac{1 - 1.0325^{-17}}{0.0325}\right]$$

$$= 1040.00(0.5805892) + 45.00(12.904947)$$
$$= 603.81 + 580.72$$
$$= \$1184.53$$

5. $FV = 5\,000\,000.00$; $PMT = 5\,000\,000.00(0.0725) = 362\,500.00$; $i = 4.25\%$;
$n = 25$; $c = 2$
$$p = 1.0425^2 - 1 = 1.08680625 - 1 = 8.680625\%$$

$$\text{Purchase price} = 5\,000\,000.00\left(1.08680625^{-25}\right) + 362\,500.00\left[\frac{1 - 1.08680625^{-25}}{0.08680625}\right]$$

$$= 5\,000\,000.00(0.1247949) + 362\,500.00(10.0822822)$$
$$= 623\,975.00 + 3\,654\,827.00$$
$$= \$4\,278\,802.00$$

7. $FV = 40\,000.00(1.03) = 41\,200.00$; $PMT = 40\,000.00(0.02) = 800.00$; $n = 30$;
$i = 3.4\%$; $c = \dfrac{2}{4} = \dfrac{1}{2}$
$$p = 1.034^{\frac{1}{2}} - 1 = 1.0168579 - 1 = 1.68579\%$$

$$\text{Purchase price} = 41\,200.00\left(1.0168579^{-30}\right) + 800.00\left[\frac{1 - 1.0168579^{-30}}{0.0168579}\right]$$

$$= 41\,200.00(0.6056085) + 800.00(23.395053)$$
$$= 24\,951.07 + 18\,716.04$$
$$= \$43\,667.11$$

9. (a) $FV = 25\,000.00$; $PMT = 25\,000.00(0.05) = 1250.00$; $i = 3.8\%$
    The interest date preceding the purchase date is June 1, 2000. Time period June 1, 2000 to December 1, 2011 is 11.5 years: $n = 23$.
    Purchase price on June 1, 2000

$$= 25\,000.00\left(1.038^{-23}\right) + 1250.00\left[\frac{1 - 1.038^{-23}}{0.038}\right]$$
$$= 25\,000.00(0.4240928) + 1250.00(15.155453)$$
$$= 10\,602.32 + 18\,944.32$$
$$= \$29\,546.64$$

    Number of days in the interest payment interval June 1, 2000 to December 1, 2000 is 183; number of days from June 1, 2000 to September 25, 2000 is 116.
    Accumulated value on September 25, 2000:
$$PV = 29\,546.64; \ r = 3.8\%; \ t = \frac{116}{183}$$

$$\text{Flat price} = 29\,546.64\left(1 + 0.038 \times \frac{116}{183}\right)$$
$$= 29\,546.64(1.0240874)$$
$$= \$30\,258.34$$

(b)   Accrued interest $= 25\,000.00(0.05)\left(\dfrac{116}{183}\right) = \$792.35$

(c)   Quoted price $= 30\,258.34 - 792.35 = \$29\,465.99$

11. $FV = 5000.00(1.04) = 5200.00$; $PMT = 5000.00(0.0475) = 237.50$
The interest date preceding the date of purchase is February 1, 2001.
The time period February 1, 2001 to August 1, 2010 is 9.5 years: $n = 19$; $i = 4.25\%$.
Purchase price on February 1, 2001

$$= 5200.00\left(1.0425^{-19}\right) + 237.50\left[\dfrac{1 - 1.0425^{-19}}{0.0425}\right]$$

$$= 5200.00(0.4534765) + 237.50(12.859376)$$

$$= 2358.08 + 3054.10$$

$$= \$5412.18$$

The number of days in the interest payment interval February 1, 2001 to August 1, 2001 is 181 days. The number of days from February 1, 2001 to May 10, 2001 is 98.
Accumulated value on May 10, 2001:

$PV = 5412.18$; $r = 4.25\%$; $t = \dfrac{98}{181}$

$$\text{Flat price} = 5412.18\left(1 + 0.0425 \times \dfrac{98}{181}\right)$$

$$= 5412.18(1.023011)$$

$$= \$5536.72$$

Accrued interest $= 5000.00(0.0475)\left(\dfrac{98}{181}\right) = \$128.59$

Quoted price $= 5536.72 - 128.59 = \$5408.13$

Exercise 15.2

A.   1.   (a)   Face value $= 25\,000.00$; $b = 3\%$
Redemption price $= 25\,000.00$; $i = 4.5\%$; $n = 20$
Since $b < i$, the bond will sell at a discount.

$$\text{Discount} = \left[25\,000.00(0.03) - 25\,000.00(0.045)\right]\left[\dfrac{1 - 1.045^{-20}}{0.045}\right]$$

$$= (25\,000.00)(-0.015)(13.007936)$$

$$= -\$4877.98$$

(b)   Purchase price $= 25\,000.00 - 4877.98 = \$20\,122.02$

3. (a) Face value = 10 000.00; $b = 4.5\%$
        Redemption price = 10 000.00(1.04) = 10 400.00; $i = 5.5\%$; $n = 30$
        Since $b < i$, the bond is expected to sell at a discount.

$$\text{Discount} = \left[10\,000.00(0.045) - 10\,400.00(0.055)\right]\left[\frac{1 - 1.055^{-30}}{0.055}\right]$$

$$= (450.00 - 572.00)(14.533745)$$
$$= (-122.00)(14.533745)$$
$$= -\$1773.12$$

    (b) Purchase price = 10 400.00 - 1773.12 = $8626.88

5. (a) Face value = 60 000.00; $b = 3.5\%$
        Redemption price = 60 000.00(1.08) = 64 800.00; $i = 4.75\%$; $n = 14$
        Since $b < i$, the bond is expected to sell at a discount.

$$\text{Discount} = \left[60\,000.00(0.035) - 64\,800.00(0.0475)\right]\left[\frac{1 - 1.0475^{-14}}{0.0475}\right]$$

$$= (2100.00 - 3078.00)(10.058778)$$
$$= (-978.00)(10.058778)$$
$$= -\$9837.48$$

    (b) Purchase price = 64 800.00 - 9837.48 = $54 962.52

7. (a) Face value = 1000.00; $b = 3.25\%$
        Redemption price = 1000.00(1.05) = 1050.00; $i = 4\%$
        The interest date preceding the date of purchase is 7 years before maturity: $n = 14$.
        Since $b < i$, the bond is expected to sell at a discount on the interest date.

$$\text{Discount} = \left[1000.00(0.0325) - 1050.00(0.04)\right]\left[\frac{1 - 1.04^{-14}}{0.04}\right]$$

$$= (32.50 - 42.00)(10.563123)$$
$$= (-9.50)(10.563123)$$
$$= -\$100.35$$

    (b) Purchase price 2 months after interest date:

$$PV = 1050.00 - 100.35 = 949.65; \quad r = 8\%; \quad t = \frac{2}{12} = \frac{1}{6}$$

$$\text{Purchase price} = 949.65\left(1 + 0.08 \times \frac{1}{6}\right)$$
$$= 949.65(1.0133333)$$
$$= \$962.31$$

B. 1. Face value = 100 000.00; $b = 2\%$
        Redemption price = 100 000.00; $i = 1.625\%$
        Since $b > i$, the bond is expected to sell at a premium.

   (a) $n = 60$

   $$\text{Premium} = \left[100\,000.00(0.02) - 100\,000.00(0.01625)\right]\left[\frac{1 - 1.01625^{-60}}{0.01625}\right]$$

   $$= 100\,000.00(0.00375)(38.143997)$$
   $$= \$14\,304.00$$

   Purchase price = 100 000.00 + 14 304.00 = $114 304.00

   (b) $n = 20$

   $$\text{Premium} = \left[100\,000.00(0.02) - 100\,000.00(0.01625)\right]\left[\frac{1 - 1.01625^{-20}}{0.01625}\right]$$

   $$= 100\,000.00(0.00375)(16.958934)$$
   $$= \$6359.60$$

   Purchase price = 100 000.00 + 6359.60 = $106 359.60

3. Face value = 25 000.00; $b = 9\%$
    Redemption price = 25 000.00; $n = 6$

   (a) $i = 13.5\%$
        Since $b < i$, the bond will sell at a discount.

   $$\text{Discount} = \left[25\,000.00(0.09) - 25\,000.00(0.135)\right]\left[\frac{1 - 1.135^{-6}}{0.135}\right]$$

   $$= 25\,000.00(-0.045)(3.9425046)$$
   $$= -\$4435.32$$

   Purchase price = 25 000.00 − 4435.32 = $20 564.68

   (b) $i = 6\%$
        Since $b > i$, the bond will sell at a premium.

   $$\text{Premium} = \left[25\,000.00(0.09) - 25\,000.00(0.06)\right]\left[\frac{1 - 1.06^{-6}}{0.06}\right]$$

   $$= 25\,000.00(0.03)(4.9173243)$$
   $$= \$3687.99$$

   Purchase price = 25 000.00 + 3687.99 = $28 687.99

5. (a) Face value $= 12(1000.00) = 12\,000.00;\ b = 5\%$
        Redemption price $= 12\,000.00;\ i = 3.5\%$
        The interest date preceding the date of purchase is March 1, 2001.
        The period March 1, 2001 to September 1, 2006 contains 5.5 years: $n = 11$.
        Since $b > i$, the premium on March 1, 2001

    $$= \left[12\,000.00(0.05) - 12\,000.00(0.035)\right]\left[\frac{1 - 1.035^{-11}}{0.035}\right]$$

    $$= 12\,000.00(0.015)(9.0015510)$$

    $$= \$1620.28$$

    (b) Purchase price on March 1, 2001 $= 12\,000.00 + 1620.28 = \$13\,620.28$
        The period March 1, 2001 to June 18, 2001 is 109 days.
        The period March 1, 2001 to September 1, 2001 is 184 days.

        $PV = 13\,620.28;\ r = 3.5\%;\ t = \dfrac{109}{184}$

        Accumulated value on June 18, 2001:

        $$\text{Purchase price} = 13\,620.28\left(1 + 0.035 \times \frac{109}{184}\right)$$

        $$= 13\,620.28(1.0207337)$$

        $$= \$13\,902.68$$

    (c) Interest accrued $= 12\,000.00(0.05)\left(\dfrac{109}{184}\right) = \$355.43$

        Quoted price $= 13\,902.68 - 355.43 = \$13\,547.25$

7. Face value $= 5\,000\,000.00;\ b = 3.625\%$

    Redemption price $= 5\,000\,000.00;\ i = 0.7\%;\ c = \dfrac{12}{2} = 6;\ n = 20$

    $p = 1.007^6 - 1 = 1.0427419 - 1 = 4.27419\%$
    Since $b < p$, discount is expected.

    $$\text{Discount} = \left[5\,000\,000.00(0.03625) - 5\,000\,000.00(0.0427419)\right]\left[\frac{1 - 1.0427419^{-20}}{0.0427419}\right]$$

    $$= 5\,000\,000.00(-0.0064919)(13.266240)$$

    $$= -\$430\,615.52$$

    Purchase price $= \$4\,569\,384.48$

Exercise **15.3**

A. 1. (a) Face value = 5000.00; $b = 3\%$
Redemption price = 5000.00; $i = 3.25\%$; $n = 7$
Since $b < i$, discount is expected.

$$\text{Discount} = \left[5000.00(0.03) - 5000.00(0.0325)\right]\left[\frac{1 - 1.0325^{-7}}{0.0325}\right]$$

$$= 5000.00(-0.0025)(6.1720)$$

$$= -\$77.15$$

Purchase price = 5000.00 − 77.15 = $4922.85

(b)

		Schedule of Accumulation of Discount			
Payment interval	*Coupon b = 3%*	*Interest on book i = 3.25%*	*Discount accumulated*	*Book value*	*Discount balance*
0				4922.85	77.15
1	150.00	159.99	9.99	4932.84	67.16
2	150.00	160.32	10.32	4943.16	56.84
3	150.00	160.65	10.65	4953.81	46.19
4	150.00	161.00	11.00	4964.81	35.19
5	150.00	161.36	11.36	4976.17	23.83
6	150.00	161.73	11.73	4987.90	12.10
7	150.00	162.10	12.10	5000.00	—
Total	1050.00	1127.15	77.15		

3. (a) Face value = 1000.00; $b = 6\%$
Redemption price = 1000.00(1.03) = 1030.00; $i = 5\%$
March 1, 2002 to September 1, 2005 is 3.5 years; $n = 7$.
Since $b > i$, premium is expected.

$$\text{Premium} = \left[1000.00(0.06) - 1030.00(0.05)\right]\left[\frac{1 - 1.05^{-7}}{0.05}\right]$$

$$= (60.00 - 51.50)(5.7863734)$$

$$= \$49.18$$

Purchase price = 1030.00 + 49.18 = $1079.18

(b)

		Interest on			
Payment interval	Coupon $b=6\%$	book $i=5\%$	Premium amortized	Book value	Premium balance
0				1079.18	49.18
1	60.00	53.96	6.04	1073.14	43.14
2	60.00	53.66	6.34	1066.80	36.80
3	60.00	53.34	6.66	1060.14	30.14
4	60.00	53.01	6.99	1053.15	23.15
5	60.00	52.66	7.34	1045.81	15.81
6	60.00	52.29	7.71	1038.10	8.10
7	60.00	51.90	8.10	1030.00	—
Total	420.00	370.82	49.18		

Schedule of Amortization of Premium

B. 1. Proceeds = 25 000.00(0.9925) = \$24 812.50
Face value = 25 000.00; b = 5.25%
Redemption price = 25 000.00; i = 6%; n = 8
$b < i \rightarrow$ discount

$$\text{Discount} = \left[25\,000.00(0.0525) - 25\,000.00(0.06)\right]\left[\frac{1 - 1.06^{-8}}{0.06}\right]$$

$$= 25\,000.00(-0.0075)(6.2097938)$$

$$= -\$1164.34$$

Book value = 25 000.00 − 1164.34 = \$23 835.66
Gain on sale = 24 812.50 − 23 835.66 = \$976.84

3. Face value = 7(1000.00) = 7000.00; b = 9.25%
Redemption price = 7000.00(1.07) = 7490.00; i = 8.25%; n = 3
$b > i \rightarrow$ premium

$$\text{Premium} = \left[7000.00(0.0925) - 7490.00(0.0825)\right]\left[\frac{1 - 1.0825^{-3}}{0.0825}\right]$$

$$= (647.50 - 617.925)(2.5655159)$$

$$= \$75.88$$

Book value = 7490.00 + 75.88   = \$7565.88
Proceeds = 7000.00(0.945)   = 6615.00
　　Loss on sale   \$ 950.88

5. Face value = 5000.00; $b = 4\%$
Redemption price = 5000.00; $i = 4.5\%$
The interest date preceding the selling date is June 1, 2002.
The time from June 1, 2002 to June 1, 2012 is 10 years: $n = 20$.
$b < i \rightarrow$ discount
Discount on June 1, 2002

$$= \left[5000.00(0.04) - 5000.00(0.045)\right]\left[\frac{1 - 1.045^{-20}}{0.045}\right]$$

$$= -25.00(15.0079364)$$

$$= -\$325.20$$

Book value on June 1, 2002 = 5000.00 − 325.20 = $4674.80
Accumulated value on September 22, 2002:
The period June 1, 2002 to September 22, 2002 is 113 days.
The period June 1, 2002 to December 1, 2002 is 183 days.
Accumulated book value on September 22, 2002

$$= 4674.80\left(1 + 0.045 \times \frac{113}{183}\right)$$

$$= 4674.80(1.0277869) = \$4804.70$$

Accrued interest on September 22, 2002

$$= 5000.00(0.04)\left(\frac{113}{183}\right) = \$123.50$$

Proceeds = 5000.00(1.01375) + 123.50

$$= 5068.75 + 123.50 = \$5192.25$$

Gain on sale = 5192.25 − 4804.70 = $387.55

Exercise 15.4

A. 1. Quoted price (initial book value) = 10 000.00(1.01375) = $10 137.50
Redemption price = $10 000.00

Average book value = $\frac{1}{2}$(10 137.50 + 10 000.00) = $10 068.75

The semi-annual interest = 10 000.00(0.03) = $300.00
The number of interest payments to maturity = 30
The total interest payments = 30(300.00) = $9 000.00
The bond premium = 10 137.50 − 10 000.00 = $137.50
Average income per interest payment interval

$$= \frac{1}{30}(9\,000.00 - 137.50) = \frac{1}{30}(8862.50) = \$295.42$$

Approximate value of $i = \dfrac{295.42}{10\,068.75} = 2.934\%$

The approximate yield rate = 2(2.934%) = 5.868%

3. Initial book value = 25 000.00(0.97125) = \$24 281.25
    Redemption price = 25 000.00(1.04) = \$26 000.00

    Average book value = $\frac{1}{2}$(24 281.25 + 26 000.00) = \$25 140.63

    Semi-annual coupon = 25 000.00(0.0375) = \$937.50
    Number of coupons to maturity = 20
    Total value of coupons = 20(937.50)) = \$18 750.00
    The bond discount = 26 000.00 − 24 281.25 = \$1718.75
    Average income per interest payment interval

    $= \frac{1}{20}$(18 750.00 + 1718.75) $= \frac{1}{20}$(20 468.75) = \$1023.44

    Approximate value of i = $\frac{1023.44}{25\,140.63}$ = 4.07086%

    Approximate yield rate = 2(4.07086%) = 8.14%

5. Initial book value = 50 000.00(0.98875) = \$49 437.50
    Redemption price = \$50 000.00

    Average book value = $\frac{1}{2}$(50 000.00 + 49 437.50) = \$49 718.75

    Nearest interest payment date is 5.5 years before maturity.
    Number of interest payments = 11
    Semi-annual interest = 50 000.00(0.045) = \$2250.00
    Total interest = 11(2250.00) = \$24 750.00
    Bond discount = 50 000.00 − 49 437.50 = \$562.50
    Average income per interest payment interval

    $= \frac{1}{11}$(24 750.00 + 562.50) $= \frac{1}{11}$(25 312.50) = \$2301.14

    Approximate value of i = $\frac{2301.14}{49\,718.75}$ = 4.62831%

    Approximate yield rate = 2(4.62831%) = 9.26%

Exercise **15.5**

A. 1. (a) FV_n = 15 000.00; i = 3.0%; n = 20

        $$15\,000.00 = PMT\left[\frac{1.03^{20} - 1}{0.03}\right]$$

        $15\,000.00 = 26.8703745PMT$
        $PMT = \$558.24$

    (b) PMT = 558.24; i = 3.0%; n = 10

        $$FV_{10} = 558.24\left[\frac{1.03^{10} - 1}{0.03}\right]$$

        $= 558.24(11.4638793)$
        $= \$6399.60$

3.   (a) $FV_n(due) = 8400.00$; $i = 0.75\%$; $n = 180$

$$8400.00 = PMT(1.0075)\left[\frac{1.0075^{180} - 1}{0.0075}\right]$$

$$8400.00 = PMT(1.0075)(378.40577)$$

$$PMT = \$22.03$$

  (b) $FV_{96}(due) = 22.03(1.0075)\left[\dfrac{1.0075^{96} - 1}{0.0075}\right]$

$$= 22.03(1.0075)(139.85616)$$

$$= \$3104.14$$

5.   (a) $FV_n = 45\,000.00$; $i = 2.5\%$; $n = 48$

$$45\,000.00 = PMT\left[\frac{1.025^{48} - 1}{0.025}\right]$$

$$45\,000.00 = 90.8595824\,PMT$$

$$PMT = \$495.27$$

  (b) $FV_{16} = 495.27\left[\dfrac{1.025^{16} - 1}{0.025}\right]$

$$= 495.27(19.3802248)$$

$$= \$9598.44$$

B.   1.   (a) $PV = 20\,000.00$; $i = 2.5\%$
                Quarterly interest $= 20\,000.00(0.025) = \$500.00$

      (b) $FV_n = 20\,000.00$; $i = 3\%$; $n = 40$

$$20\,000.00 = PMT\left[\frac{1.03^{40} - 1}{0.03}\right]$$

$$20\,000.00 = 75.401260\,PMT$$

$$PMT = \$265.25$$

(c)   Quarterly cost $= 500.00 + 265.25 = \$765.25$

(d) $FV_{24} = 265.25\left[\dfrac{1.03^{24} - 1}{0.03}\right]$

$\quad\quad\quad = 265.25(34.42647)$

$\quad\quad\quad = \$9131.62$

Book value after 6 years $= 20\,000.00 - 9131.62 = \$10\,868.38$

3.   (a) $PV = 10\,000.00;\ i = 0.625\%$
Monthly interest $= 10\,000.00(0.00625) = \$62.50$

(b) $FV_n = 10\,000.00;\ i = 0.5\%;\ n = 60$

$10\,000.00 = PMT\left[\dfrac{1.005^{60} - 1}{0.005}\right]$

$10\,000.00 = 69.7700305 PMT$

$\quad PMT = \$143.33$

(c)   Monthly cost $= 62.50 + 143.33 = \$205.83$

(d) $FV_{48} = 143.33\left[\dfrac{1.005^{48} - 1}{0.005}\right]$

$\quad\quad\quad = 143.33(54.097832)$

$\quad\quad\quad = \$7753.84$

Book value after 4 years $= 10\,000.00 - 7753.84 = \$2246.16$

5.   (a) $PV = 95\,000.00;\ i = 4.5\%$
Semi-annual interest $= 95\,000.00(0.045) = \$4275.00$

(b) $FV_n = 95\,000.00;\ i = 3.5\%;\ n = 40$

$95\,000.00 = PMT\left[\dfrac{1.035^{40} - 1}{0.035}\right]$

$95\,000.00 = 84.5502778 PMT$

$\quad PMT = \$1123.59$

(c)   Semi-annual cost = 4275.00 + 1123.59 = $5398.59

(d) $FV_{30} = 1123.59\left[\dfrac{1.035^{30} - 1}{0.035}\right]$

$\qquad\qquad = 1123.59(51.6226773)$

$\qquad\qquad = \$58\,002.72$

Book value after 15 years = 95 000.00 − 58 002.72 = $36 997.28

C.   1.   (a) $FV_n = 75\,000.00;\ i = 1.25\%;\ n = 24$

$75\,000.00 = PMT\left[\dfrac{1.0125^{24} - 1}{0.0125}\right]$

$75\,000.00 = 27.788084\,PMT$

$\qquad PMT = \$2699.00$

(b)   Total paid = 2699.00(24) = $64 776.00

(c)   Interest = 75 000.00 − 64 776.00 = $10 224.00

3. $FV_n = 20\,000.00;\ i = 5.5\%;\ n = 7$

$20\,000.00 = PMT\left[\dfrac{1.055^{7} - 1}{0.055}\right]$

$20\,000.00 = 8.2668938\,PMT$

$\qquad PMT = \$2419.29$

Sinking Fund Schedule

Payment number	Periodic payment	Interest earned	Increase in fund	Balance after payment
0				—
1	2 419.29	—	2 419.29	2 419.29
2	2 419.29	133.06	2 552.35	4 971.64
3	2 419.29	273.44	2 692.73	7 664.37
4	2 419.29	421.54	2 840.83	10 505.20
5	2 419.29	577.79	2 997.08	13 502.28
6	2 419.29	742.63	3 161.92	16 664.20
7	2 419.29	916.53	3 335.82	20 000.02
Total	16 935.03	3064.99	20 000.02	

5. PMT = 2419.29; i = 5.5%; n = 3

$$FV_n = 2419.29\left[\frac{1.055^3 - 1}{0.055}\right]$$

$$= 2419.29(3.1680250)$$

$$= \$7664.37$$

Interest in Year 4 = 7664.37(0.055) = \$421.54
Increase in Fund = 2419.29 + 421.54 = \$2840.83

7. (a) FV_n = 100 000.00; i = 0.625%; n = 180

$$100\,000.00 = PMT\left[\frac{1.00625^{180} - 1}{0.00625}\right]$$

$$100\,000.00 = 331.11228PMT$$

$$PMT = \$302.01$$

(b) Balance after 5 years:

$$FV_{60} = 302.01\left[\frac{1.00625^{60} - 1}{0.00625}\right]$$

$$= 302.01(72.527105)$$

$$= \$21\,903.91$$

(c) Balance after 99 intervals:

$$FV_{99} = 302.01\left[\frac{1.00625^{99} - 1}{0.00625}\right]$$

$$= 302.01(136.48548)$$

$$= \$41\,219.98$$

Interest in 100th interval = 41 219.98(0.00625) = \$257.62

(d) Balance after 149 intervals:

$$FV_{149} = 302.01\left[\frac{1.00625^{149} - 1}{0.00625}\right]$$

$$= 302.01(244.85368)$$

$$= \$73\,948.26$$

Interest in interval 150 = 73 948.26(0.00625) = \$462.18
Increase in fund = 302.01 + 462.18 = \$764.19

(e) Last 3 payments are PMT_{178}, PMT_{179}, PMT_{180}.

Balance after 177 payments:

$$FV_{177} = 302.01\left[\frac{1.00625^{177} - 1}{0.00625}\right]$$

$$= 302.01(322.01784)$$

$$= \$97\,252.61$$

Partial Sinking Fund Schedule

Payment interval	Periodic payment	Interest earned	Increase in fund	Balance at end
0				—
1	302.01	—	302.01	302.01
2	302.01	1.89	303.90	605.91
3	302.01	3.79	305.80	911.71
⋮	⋮	⋮	⋮	⋮
⋮	⋮	⋮	⋮	⋮
177	⋮	⋮	⋮	97 252.61
178	302.01	607.83	909.84	98 162.45
179	302.01	613.52	915.53	99 077.98
180	302.01	619.24	921.25	99 999.23
Total	54 361.80	45 637.43	99 999.23	

9. (a) PV = 300 000.00; $i = 8.25\%$

Annual interest = 300 000.00(0.0825) = \$24 750.00

(b) $FV_n = 300\,000.00$; $i = 5.5\%$; $n = 20$

$$300\,000.00 = \text{PMT}\left[\frac{1.055^{20} - 1}{0.055}\right]$$

$$300\,000.00 = 34.868318\text{PMT}$$

$$\text{PMT} = \$8604.00$$

(c) Annual cost = 24 750.00 + 8604.00 = \$33 354.00

(d) Balance after 9 years:

$$FV_9 = 8604.00\left[\frac{1.055^9 - 1}{0.055}\right]$$

$$= 8604.00(11.256260)$$

$$= \$96\,849.00$$

Interest in year 10 = 96 849.00(0.055) = \$5327.00

Increase in fund = 5327.00 + 8604.00 = \$13 931.00

(e)   Balance after 15 years:

$$FV_{15} = 8604.00\left[\frac{1.055^{15} - 1}{0.055}\right]$$

$$= 8604.00(22.4086635)$$

$$= \$192\,804.00$$

Book value of debt $= 300\,000.00 - 192\,804.00 = \$107\,196.00$

(f)   Last 3 payments are $PMT_{18}, PMT_{19}, PMT_{20}$.

Balance after 17 years:

$$FV_{17} = 8604.00\left[\frac{1.055^{17} - 1}{0.055}\right]$$

$$= 8604.00(26.996403)$$

$$= \$232\,277.00$$

Partial Sinking Fund Schedule

Payment interval	Periodic payment	Interest earned	Increase in fund	Balance in fund	Book value of debt
0					300 000.00
1	8 604.00	—	8 604.00	8 604.00	291 396.00
2	8 604.00	473.00	9 077.00	17 681.00	282 319.00
3	8 604.00	972.00	9 576.00	27 257.00	272 743.00
⋮	⋮	⋮	⋮	⋮	⋮
⋮	⋮	⋮	⋮	⋮	⋮
17	⋮	⋮	⋮	232 277.00	67 723.00
18	8 604.00	12 775.00	21 379.00	253 656.00	46 344.00
19	8 604.00	13 951.00	22 555.00	276 211.00	23 789.00
20	8 597.00	15 192.00	23 789.00	300 000.00	—
Total	172 073.00	127 927.00	300 000.00		

Review Exercise

1. $FV = 5000.00; \quad PMT = 5000.00(0.0575) = 287.50; \quad n = 24$

(a) $i = 5.25\%$

$$PP = 5000.00\left(1.0525^{-24}\right) + 287.50\left[\frac{1 - 1.0525^{-24}}{0.0525}\right]$$

$$= 5000.00(0.2928664) + 287.50(13.469212)$$

$$= 1464.33 + 3872.40$$

$$= \$5336.73$$

(b) $i = 6.5\%$

$$PP = 5000.00\left(1.065^{-24}\right) + 287.50\left[\frac{1 - 1.065^{-24}}{0.065}\right]$$

$$= 5000.00(0.220602) + 287.5(11.990739)$$

$$= 1103.01 + 3447.34$$

$$= \$4550.35$$

3. $FV = 25\,000.00(1.04) = 26\,000.00$; $PMT = 25\,000.00(0.0225) = 562.50$;

$i = 8.25\%$; $c = \dfrac{1}{4}$; $n = 24$

$p = 1.0825^{\frac{1}{4}} - 1 = 1.020016 - 1 = 2.0016\%$

$$PP = 26\,000.00\left(1.020016^{-24}\right) + 562.50\left[\frac{1 - 1.020016^{-24}}{0.020016}\right]$$

$$= 26\,000.00(0.6214875) + 562.50(18.910498)$$

$$= 16\,158.67 + 10\,637.16$$

$$= \$26\,795.83$$

5. (a) Face value $= 4(5000.00) = 20\,000.00$; $b = 3.5\%$
Redemption price $= 20\,000.00$; $i = 3\%$; $n = 14$
$b > i \rightarrow$ premium

$$Premium = \left[20\,000.00(0.035) - 20\,000.00(0.03)\right]\left[\frac{1 - 1.03^{-14}}{0.03}\right]$$

$$= (20\,000.00)(0.005)(11.296073)$$

$$= \$1129.61$$

Purchase price $= 20\,000.00 + 1129.61 = \$21\,129.61$

(b) Face value $= 20\,000.00$; $b = 3.5\%$
Redemption price $= 4(5000.00)(1.07) = 21\,400.00$; $i = 3\%$; $n = 14$
$b > i \rightarrow$ premium expected

$$Premium = \left[20\,000.00(0.035) - 21\,400.00(0.03)\right]\left[\frac{1 - 1.03^{-14}}{0.03}\right]$$

$$= (700.00 - 642.00)(11.296073)$$

$$= \$655.17$$

Purchase price $= 21\,400.00 + 655.17 = \$22\,055.17$

7. (a) Face value = 100 000.00; b = 2.5%
Redemption price = 100 000.00; i = 3.5%
The interest dates are July 15 and January 15.
The interest date preceding the purchase date is January 15, 2001.
The period January 15, 2001 to July 15, 2012 is 11.5 years; n = 23.
$b < i \rightarrow$ discount
Discount on January 15, 2001

$$= \left[100\,000.00(0.025 - 0.035)\right]\left[\frac{1 - 1.035^{-23}}{0.035}\right]$$

$$= 100\,000.00(-0.01)(15.620410)$$

$$= -\$15\,620.41$$

(b) Purchase price on January 15, 2001 = 100 000.00 − 15 620.41 = \$84 379.59
The interest payment interval January 15, 2001 to July 15, 2001 is 181 days.
The interest period January 15, 2001 to April 18, 2001 is 93 days.
Accumulated value on April 18, 2001

$$= 84\,379.59\left(1 + 0.035 \times \frac{93}{181}\right)$$

$$= 84\,379.59(1.0179834)$$

$$= \$85\,897.02$$

(c) Accrued interest $= 100\,000.00(0.025)\left(\frac{93}{181}\right) = \$1284.53$
Quoted price = 85 897.02 − 1284.53 = \$84 612.49

9. FV = 5000.00(1.08) = 5400.00; PMT = 5000.00(0.04) = 200.00;
i = 5%; n = 20

$$\text{Purchase price} = 5400.00\left(1.05^{-20}\right) + 200.00\left[\frac{1 - 1.05^{-20}}{0.05}\right]$$

$$= 5400.00(0.3768895) + 200.00(12.4622103)$$

$$= 2035.20 + 2492.44$$

$$= \$4527.64$$

11.   (a)   Face value $= 25\,000.00$; $b = 6.5\%$
Redemption price $= 25\,000.00(1.07) = 26\,750.00$; $i = 7.25\%$
Interest dates are June 15 and December 15.
The interest date preceding the purchase date is December 15, 2002.
The time interval December 15, 2002 to June 15, 2014 is 11.5 years: $n = 23$.
$b < i \rightarrow$ discount
Discount on December 15, 2002

$$= \left[25\,000.00(0.065) - 26\,750.00(0.0725)\right]\left[\frac{1 - 1.0725^{-23}}{0.0725}\right]$$

$$= (1625.00 - 1939.375)(11.035549)$$
$$= (-314.375)(11.035549)$$
$$= -\$3469.30$$

(b)   Purchase price on December 15, 2002 $= 26\,750.00 - 3469.30 = \$23\,280.70$
The payment interval December 15, 2002 to June 15, 2003 is 182 days.
The interest period December 15, 2002 to May 9, 2003 is 145 days.
Accumulated value on May 9, 2003

$$= 23\,280.70\left(1 + 0.0725 \times \frac{145}{182}\right)$$

$$= 23\,280.70(1.057761)$$
$$= \$24\,625.42$$

(c)   Accrued interest on May 9, 2003

$$= 25\,000.00(0.065)\left(\frac{145}{182}\right) = \$1294.64$$

Quoted price $= 24\,625.42 - 1294.64 = \$23\,330.78$

13. Face value $= 1000.00$; $b = 8.5\%$
     Redemption price $= 1000.00$; $i = 10.5\%$; $n = 6$
     $b < i \rightarrow$ discount

$$\text{Discount} = \left[1000.00(0.085 - 0.105)\right]\left[\frac{1 - 1.105^{-6}}{0.105}\right]$$

$$= 1000.00(-0.020)(4.292179)$$

$$= -\$85.84$$

Purchase price $= 1000.00 - 85.84 = \$914.16$

Schedule of Accumulation of Discount

Payment interval	Coupon $b = 8.5\%$	Interest on book $i = 10.5\%$	Discount accumulated	Book value	Discount balance
0				914.16	85.84
1	85.00	95.99	10.99	925.15	74.85
2	85.00	97.14	12.14	937.29	62.71
3	85.00	98.42	13.42	950.71	49.29
4	85.00	99.82	14.82	965.53	34.47
5	85.00	101.38	16.38	981.91	18.09
6	85.00	103.09	18.09	1000.00	—
Total	510.00	595.84	85.84		

15. Face value $= 20\,000.00$; $b = 4.75\%$
     Redemption price $= 20\,000.00(1.05) = 21\,000.00$; $i = 4\%$; $n = 6$
     $b > i \rightarrow$ premium expected

$$\text{Premium} = \left[20\,000.00(0.0475) - 21\,000.00(0.04)\right]\left[\frac{1 - 1.04^{-6}}{0.04}\right]$$

$$= (950.00 - 840.00)(5.2421369)$$

$$= \$576.64$$

Purchase price $= 21\,000.00 + 576.64 = \$21\,576.64$

Schedule of Amortization of Premium

Payment interval	Coupon $b = 4.75\%$	Interest on book $i = 4\%$	Premium amortized	Book value	Premium balance
0				21 576.64	576.64
1	950.00	863.07	86.93	21 489.71	489.71
2	950.00	859.59	90.41	21 399.30	399.30
3	950.00	855.97	94.03	21 305.27	305.27
4	950.00	852.21	97.79	21 207.48	207.48
5	950.00	848.30	101.70	21 105.78	105.78
6	950.00	844.22	105.78	21 000.00	—
Total	5700.00	5123.36	576.64		

17. Face value = 10 000.00; b = 1.25%
Redemption price = 10 000.00(1.06) = 10 600.00; i = 2.25%
Interest dates are November 15, February 15, May 15, August 15.
The interest date preceding the date of sale is August 15, 2002.
The time period August 15, 2002 to November 15, 2012 is 10.25 years: n = 41.
$b < i \rightarrow$ discount
Discount on August 15, 2002

$$= \left[10\,000.00(0.0125) - 10\,600.00(0.0225)\right]\left[\frac{1 - 1.0225^{-41}}{0.0225}\right]$$

$$= (125.00 - 238.50)(26.595132)$$

$$= -\$3018.55$$

Book value on August 15, 2002 = 10 600.00 − 3018.55 = \$7581.45
The interest payment interval August 15, 2002 to November 15, 2002 is 92 days.
The interest period August 15, 2002 to September 10, 2002 is 26 days.
Accumulated value on September 10, 2002

$$= 7581.45\left(1 + 0.0225 \times \frac{26}{92}\right)$$

$$= 7581.45(1.0063587)$$

$$= \$7629.66$$

Proceeds = 10 000.00(0.9275) = \$9275.00
Gain on sale = 9275.00 − 7629.66 = \$1645.34

19. Quoted price = 10 000.00(0.9875) = \$9875.00
Redemption price = 10 000.00(1.02) = \$10 200.00

Average book value = $\frac{1}{2}$(10 200.00 + 9875.00) = \$10 037.50

Interest payment dates are October 15, January 15, April 15, July 15.
The interest date preceding the date of purchase is April 15, 2001.
The time period April 15, 2001 to October 15, 2013 is 12.5 years.
The number of interest payments to maturity is 50.
Quarterly interest payment = 10 000.00(0.01875) = \$187.50
Total interest = 50(187.50) = \$9375.00
Bond discount = 10 200.00 − 9875.00 = \$325.00
Average income per interest period

$$= \frac{1}{50}(9375.00 + 325.00) = \frac{1}{50}(9700.00) = \$194.00$$

Approximate value of $i = \dfrac{194.00}{10\,037.50} = 1.93275\%$

The approximate yield rate = 4(1.93275%) = 7.73%

21. (a) Face value = 50 000.00; b = 1.625%
Redemption price = 50 000.00; i = 2%; n = 48
Since $b < i \rightarrow$ discount

$$\text{Discount} = \left[50\,000.00(0.01625 - 0.02)\right]\left[\frac{1 - 1.02^{-48}}{0.02}\right]$$

$$= 50\,000.00(-0.00375)(30.673120)$$

$$= -\$5751.21$$

Purchase price = 50 000.00 − 5751.21 = \$44 248.79

(b) After nine years, $n = (12 - 9)(4) = 12$

$$\text{Discount} = 50\,000.00(-0.00375)\left[\frac{1 - 1.02^{-12}}{0.02}\right]$$

$$= 50\,000.00(-0.00375)(10.575341)$$

$$= -\$1982.88$$

Book value $= 50\,000.00 - 1982.88 = \$48\,017.12$

(c) Proceeds from sale of bond $= 50\,000.00(0.99625) = \$49\,812.50$
Gain on sale $= 49\,812.50 - 48\,017.12 = \$1795.38$

23. Quoted price $= 5000.00(0.955) = \$4775.00$
Redemption price $= \$5000.00$

Average book value $= \frac{1}{2}(5000.00 + 4775.00) = \$4887.50$

Interest dates are August 1, February 1.
The interest date closest to the date of purchase is February 1, 2003.
The time interval February 1, 2003 to August 1, 2014 is 11.5 years.
The number of interest payments is 23.
Semi-annual interest $= 5000.00(0.0725) = \$362.50$
Total interest $= 23(362.50) = \$8337.50$
Bond discount $= 5000.00 - 4775.00 = \$225.00$
Average income per interest interval

$$= \frac{1}{23}(8337.50 + 225.00) = \frac{1}{23}(8562.50) = \$372.28$$

Approximate value of $i = \frac{372.28}{4887.50} = 7.62\%$

The approximate yield rate $= 2(7.62\%) = 15.24\%$

25. (a) $FV_n = 110\,000.00;\ i = 3.75\%;\ n = 10$

$$110\,000.00 = PMT\left[\frac{1.0375^{10} - 1}{0.0375}\right]$$

$$110\,000.00 = 11.867838PMT$$

$$PMT = \$9268.75$$

(b) $$FV_3 = 9268.75\left[\frac{1.0375^3 - 1}{0.0375}\right]$$

$$= 9268.75(3.1139063)$$

$$= \$28\,862.02$$

(c) $FV_5 = 9268.75 \left[\dfrac{1.0375^5 - 1}{0.0375} \right]$

   $= 9268.75(5.3893281)$

   $= \$49\,952.33$

Interest in $PMT_6 = 49\,952.33(0.0375) = \$1873.21$

(d)

Sinking Fund Schedule

Payment interval	Periodic payment	Interest earned	Increase in fund	Balance at end
0				—
1	9268.75	—	9 268.75	9 268.75
2	9268.75	347.58	9 616.33	18 885.08
3	9268.75	708.19	9 976.94	28 862.02
4	9268.75	1 082.33	10 351.08	39 213.10
5	9268.75	1 470.49	10 739.24	49 952.34
6	9268.75	1 873.21	11 141.96	61 094.30
7	9268.75	2 291.04	11 559.79	72 654.09
8	9268.75	2 724.53	11 993.28	84 647.37
9	9268.75	3 174.28	12 443.03	97 090.40
10	9268.75	3 640.89	12 909.64	110 000.04
Total	92 687.50	17 312.54	110 000.04	

27. (a) $PV = 100\,000.00$; $i = 13.75\%$

   Annual interest $= 100\,000.00(0.1375) = \$13\,750.00$

(b) $FV_n = 100\,000.00$; $i = 11.5\%$; $n = 8$

   $100\,000.00 = PMT \left[\dfrac{1.115^8 - 1}{0.115} \right]$

   $100\,000.00 = 12.077438 PMT$

   $PMT = \$8279.90$

(c) Annual cost $= 13\,750.00 + 8279.90 = \$22\,029.90$

(d) Balance after 3 years:

   $FV_3 = 8279.90 \left[\dfrac{1.115^3 - 1}{0.115} \right]$

   $= 8279.90(3.358225)$

   $= \$27\,805.77$

   Book value $= 100\,000.00 - 27\,805.77 = \$72\,194.23$

(e)   Balance after 5 years:

$$FV_5 = 8279.90\left[\frac{1.115^5 - 1}{0.115}\right]$$

$$= 8279.90(6.2900293)$$

$$= \$52\,080.81$$

Interest earned $= 52\,080.81(0.115) = \$5989.29$

(f)

Sinking Fund Schedule

Payment interval	Periodic payment	Interest earned	Increase in fund	Balance in fund	Book value of debt
0					100 000.00
1	8 279.90	—	8 279.90	8 279.90	91 720.10
2	8 279.90	952.19	9 232.09	17 511.99	82 488.01
3	8 279.90	2 013.88	10 293.78	27 805.77	72 194.23
4	8 279.90	3 197.66	11 477.56	39 283.33	60 716.67
5	8 279.90	4 517.58	12 797.48	52 080.81	47 919.19
6	8 279.90	5 989.29	14 269.19	66 350.00	33 650.00
7	8 279.90	7 630.25	15 910.15	82 260.15	17 739.85
8	8 279.93	9 459.92	17 739.85	100 000.00	—
Total	66 239.23	33 760.77	100 000.00		

29.   (a) $FV_n = 60\,000.00$; $i = 2\%$; $n = 20$

$$60\,000.00 = PMT\left[\frac{1.02^{20} - 1}{0.02}\right]$$

$$60\,000.00 = 24.297370PMT$$

$$PMT = \$2469.40$$

(b)   Total payments $= 2469.40(20) = \$49\,388.00$
Interest $= 60\,000.00 - 49\,388.00 = \$10\,612.00$

(c)   Balance after 2 years: $n = 8$

$$FV_8 = 2469.40\left[\frac{1.02^8 - 1}{0.02}\right]$$

$$= 2469.40(8.5829691)$$

$$= \$21\,194.78$$

(d)   Balance after 14th payment:

$$FV_{14'} = 2469.40\left[\frac{1.02^{14} - 1}{0.02}\right]$$

$$= 2469.40(15.973938)$$

$$= \$39\,446.04$$

Interest in 15th interval

$$= \$39\,446.04(0.02) = \$788.92$$

31.   (a) PV_n(defer) = 15 000.00; PMT = 17 500.00(due); $i = 6\%$; $d = 50 - 1 = 49$

$$15\,000.00 = 17\,500.00\left(1.06^{-49}\right)\left[\frac{1 - 1.06^{-n}}{0.06}\right]$$

$$15\,000.00 = 17\,500.00(0.05754566)\left[\frac{1 - 1.06^{-n}}{0.06}\right]$$

$$14.895004 = \frac{1 - 1.06^{-n}}{0.06}$$

$$1.06^{-n} = 0.1062998$$

$$-n\ln 1.06 = \ln 0.1062998$$

$$-0.058268908n = -2.241492$$

$$n = 38.468063$$

The number of payments is 39.

(b)   Balance at end of 38th payment interval:

$n = 38.468063 - 38 = 0.468063$

$$PV_{38} = 17\,500.00(1.06)\left[\frac{1 - 1.06^{-0.468063}}{0.06}\right]$$

$$= 17\,500.00(1.06)(0.4484159)$$

$$= \$8318.12$$

Self-Test

1.   FV = 10 000.00; PMT = 10 000.00(0.025) = 250.00

$i = 2.75\%$; $n = 60$

$$\text{Purchase price} = 10\,000.00(1.0275)^{-60} + 250.00\left[\frac{1 - 1.0275^{-60}}{0.0275}\right]$$

$$= 10\,000.00(0.1963768) + 250.00(29.222662)$$

$$= 1963.77 + 7305.67$$

$$= \$9269.44$$

3. $FV = 5000.00(1.04) = 5200.00$; $PMT = 5000.00(0.04) = 200.00$
    $i = 3.25\%$; $n = 12$
    Premium (or discount)

    $$= (5000.00 \times 0.04 - 5200.00 \times 0.0325)\left[\frac{1 - 1.0325^{-12}}{0.0325}\right]$$

    $$= (200.00 - 169.00)(9.8070764)$$

    $$= \$304.02 \text{ (premium)}$$

5. Face value $= 5000.00$; $b = 3.5\%$
    Redemption value $= 5000.00(1.02) = 5100.00$; $i = 4.25\%$
    Interest dates: December 15, June 15
    The interest date preceding the date of purchase is June 15, 2001.
    The time interval June 15, 2001 to December 15, 2012 is 11.5 years: $n = 23$.
    Premium or discount on June 15, 2001

    $$= \left[5000.00(0.035) - 5100.00(0.0425)\right]\left[\frac{1 - 1.0425^{-23}}{0.0425}\right]$$

    $$= (175.00 - 216.75)(14.495796)$$

    $$= (-41.75)(14.495796)$$

    $$= -\$605.20 \text{ (discount)}$$

    Purchase price on June 15, 2001
    $= 5100.00 - 605.20 = \$4494.80$
    The payment interval June 15, 2001 to December 15, 2001 is 183 days.
    The interest period June 15, 2001 to November 9, 2001 is 147 days.
    Accumulated value on November 9, 2001

    $$= 4494.80\left(1 + 0.0425 \times \frac{147}{183}\right)$$

    $$= 4494.80(1.0341393)$$

    $$= \$4648.25$$

    Accrued interest on November 9, 2001

    $$= 5000.00(0.035)\left(\frac{147}{183}\right) = \$140.57$$

    Quoted price $= 4648.25 - 140.57 = \$4507.68$

7. Quoted price $= 100\,000.00(1.02625) = \$102\,625.00$
Redemption value $= 100\,000.00$; $b = 6.5\%$

Average book value $= \dfrac{1}{2}(100\,000.00 + 102\,625.00) = \$101\,312.50$

Payment dates are July 15, January 15.
The interest payment date preceding purchase is July 15, 2003.
The time interval July 15, 2003 to July 15, 2010 is 7 years: $n = 14$.
Semi-annual interest $= 100\,000.00(0.065) = \$6500.00$
Total interest $= 14 \times 6500.00 = \$91\,000.00$
Premium paid $= 102\,625.00 - 100\,000.00 = \$2625.00$
Average income per interest payment interval

$\dfrac{91\,000.00 - 2625.00}{14} = \dfrac{88\,375.00}{14} = \$6312.50$

Approximate value of $i = \dfrac{6312.50}{101\,312.50} = 6.23\%$

The approximate yield rate $= 2(6.23\%) = 12.46\%$

9. Quoted price $= 10\,000.00(0.93875) = \$9387.50$
Redemption value $= \$10\,000.00$

Average book value $= \dfrac{1}{2}(10\,000.00 + 9387.50) = \$9693.75$

Interest payment dates are December 1, June 1.
The interest payment date closest to purchase is June 1, 2001.
The time interval June 1, 2001 to December 1, 2012 is 11.5 years: $n = 23$.
Semi-annual coupon $= 10\,000.00(0.06) = \$600.00$
Total interest $= 23(600.00) = \$13\,800.00$
Bond discount $= 9387.50 - 10\,000.00 = -\$612.50$
Average income per interest payment interval

$= \dfrac{1}{23}(13\,800.00 + 612.50) = \dfrac{1}{23}(14\,412.50) = \$626.63$

Approximate value of $i = \dfrac{626.63}{9693.75} = 6.46\%$
The approximate yield rate $= 2(6.46\%) = 12.92\%$

11. $FV_n(\text{due}) = 165\,000.00$; $i = 0.625\%$; $n = 72$

$165\,000.00 = PMT(1.00625)\left[\dfrac{1.00625^{72} - 1}{0.00625}\right]$

$165\,000.00 = PMT(1.00625)(90.578789)$

$\quad PMT = \$1810.30$

Balance after 19th interval:
$PMT = 1810.30$; $i = 0.625\%$; $n = 19$

$FV_n(\text{due}) = 1810.30(1.00625)\left[\dfrac{1.00625^{19} - 1}{0.00625}\right]$

$\quad = 1810.30(1.00625)(20.107566)$

$\quad = \$36\,628.23$

Interest earned in 20th payment interval
$= (36\,628.23 + 1810.30)(0.00625) = \$240.24$

13. $FV_n = 750\,000.00$; $i = 9\%$; $n = 15$

$$750\,000.00 = PMT\left[\frac{1.09^{15} - 1}{0.09}\right]$$

$750\,000.00 = 29.3609162 PMT$

$PMT = \$25\,544.16$

Balance after 5 years:

$PMT = 25\,544.16$; $i = 7\%$; $n = 5$

$$FV_n = 25\,544.16\left[\frac{1.07^5 - 1}{0.07}\right] = 25\,544.16(5.750739) = \$146\,897.80$$

Book value $= 750\,000.00 - 146\,897.80 = \$603\,102.20$

16 Investment decision applications

Exercise 16.1

A. 1. *Alternative* 1

PV of $20 000 in 3 years $= 20\,000\left(1.12^{-3}\right) = 20\,000(0.7117802) = \$14\,236$

PV of $60 000 in 6 years $= 60\,000\left(1.12^{-6}\right) = 60\,000(0.5066311) = \underline{\quad 30\,398}$

PV of Alternative 1 $= \$44\,634$

Alternative 2

PV of $13 000 at the end of each of the next six years

$$= 13\,000\left[\frac{1 - 1.12^{-6}}{0.12}\right] = 13\,000(4.1114073) = \$53\,448$$

Since PV of Alternative 2 > PV of Alternative 1, Alternative 2 is preferable at 12%.

3. PV of Alternative 1:

PV of $50 000 $= 50\,000\left(1.16^{-4}\right) = 50\,000(0.5522911) = \$27\,615$

PV of $40 000 $= 40\,000\left(1.16^{-7}\right) = 40\,000(0.3538295) = \quad 14\,153$

PV of $30 000 $= 30\,000\left(1.16^{-10}\right) = 30\,000(0.2266836) = \underline{\quad 6\,801}$

PV of Alternative 1 $= \$48\,569$

PV of Alternative 2: PMT $= 750$; $i = 16\%$; $n = 120$

$$c = \frac{1}{12},\ p = 1.16^{\frac{1}{12}} - 1 = 1.012445138 - 1 = 1.2445138\%$$

$$PV = 750\left[\frac{1 - 1.012445138^{-120}}{0.012445138}\right] = 750(62.138033) = \$46\,604$$

Since PV of Alternative 1 > PV of Alternative 2, Alternative 1 is preferred.

B. 1. PV of Alternative 1:

PV of $25 000 now $= \$25\,000$

PV of $50 000 $= 50\,000\left(1.06^{-5}\right) = 50\,000(0.7472582) = \underline{\quad 37\,363}$

PV of Alternative 1 $= \$62\,363$

PV of Alternative 2:

$$= 10\,000\left[\frac{1 - 1.06^{-10}}{0.06}\right] = 10\,000(7.360087) = \$73\,601$$

At 13%, Alternative 2 is preferable.

3. PV of buying:

PV of cash payment	=	$90 000

Less PV of 30 000 $= 30\,000 \left(1.08^{-20}\right) = 30\,000(0.2145482) =$ ___6 436___

PV of cost of buying	=	$83 564

PV of leasing:

$$PV_n(\text{due}) = 10\,000(1.08)\left[\frac{1 - 1.08^{-20}}{0.08}\right] = 10\,000(1.08)(9.8181474) = \$106\,036$$

Since PV of the cost of leasing > PV of cost of buying, the warehouse should be purchased.

Exercise 16.2

A. 1. PV of inflows:

PMT = 3500; i = 3%; n = 28

$$PV_n = 3500\left[\frac{1 - 1.03^{-28}}{0.03}\right] = 3500(18.764108) = \$65\,674$$

PV of outflows:

Immediate outlay	=	$50 000

PV of 30 000 $= 30\,000 \left(1.03^{-12}\right) = 30\,000(0.7013799) =$ ___21 041___

	=	$71 041

Net present value = 65 674 − 71 041 = −$5367
Since NPV < 0, the investment should be rejected.

3. *Alternative* 1:

$$PV_{IN} = 7000\left(1.085^{-14}\right) = 7000(0.3191418) = \$2234$$

PV_{OUT}	=	2000
NPV in Alternative 1	=	$ 234

Alternative 2:

$$PV_{IN} = 250\left[\frac{1 - 1.085^{-14}}{0.085}\right] = 250(8.0100967) = \$2003$$

PV_{OUT}	=	1800
NPV of Alternative 2	=	$ 203

Since the net present value of Alternative 1 is greater than the net present value of Alternative 2, Alternative 1 is preferred.

B. 1. *Project A*
        PV_{IN}

$$4000\left(1.12^{-4}\right) \ = \ 4000(0.6355181) \ = \ \$2542$$

$$9000\left(1.12^{-9}\right) \ = \ 9000(0.36061) \qquad = \ \underline{\ 3245}$$

$$\text{Total } PV_{IN} \qquad\qquad = \ \$5787$$

$$PV_{OUT} \qquad\qquad\qquad\qquad\quad = \ \underline{\ 4000}$$

$$NPV \qquad\qquad\qquad\qquad = \ \$1787$$

Project B

$$PV_{IN} \ = \ 1500\left[\frac{1 \ - \ 1.12^{-9}}{0.12}\right] \ = \ 1500(5.3282498) \qquad = \ \$7992$$

$$PV_{OUT} \ = \ 4000 \ + \ 2000\left(1.12^{-3}\right)$$

$$= \ 4000 \ + \ 2000(0.7117802) \ = \ 4000 \ + \ 1424 \ = \ \underline{\ 5424}$$

$$NPV \qquad\qquad\qquad\qquad\qquad\qquad = \ \$2568$$

Project B is preferred at 12%.

3. *End of year*

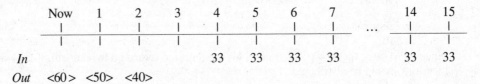

$$\text{PMT} \ = \ 33\,000; \ n \ = \ 12; \ i \ = \ 14\%; \ d \ = \ 3$$

$$PV_{IN} \ = \ 33\,000\left(1.14^{-3}\right)\left[\frac{1 \ - \ 1.14^{-12}}{0.14}\right]$$

$$= \ 33\,000(0.6749715)(5.6602921) \qquad\qquad = \ \$126\,078$$

$$PV_{OUT} \qquad 60\,000 \text{ now} \qquad\qquad\qquad \$60\,000$$

$$50\,000\left(1.14^{-1}\right) \ = \ 50\,000(0.877193) \qquad 43\,860$$

$$40\,000\left(1.14^{-2}\right) \ = \ 40\,000(0.7694675) \qquad \underline{\ 30\,779} \qquad \underline{\ 134\,639}$$

$$NPV \qquad\qquad\qquad\qquad\qquad\qquad\qquad = \ <\$\ 8\,561>$$

Since NPV < 0, the project will not return 14% on the investment and therefore should not be undertaken.

5. *End of year*

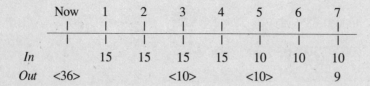

	Now	1	2	3	4	5	6	7
In		15	15	15	15	10	10	10
Out	<36>			<10>		<10>		9

PV_{IN}: $PMT = 15\,000$; $n = 4$; $i = 20\%$

$$15\,000\left[\frac{1 - 1.20^{-4}}{0.20}\right] = 15\,000(2.5887346) \qquad = \$38\,831$$

$PMT = 10\,000$; $n = 3$; $d = 4$; $i = 20\%$

$$10\,000\left(1.20^{-4}\right)\left[\frac{1 - 1.20^{-3}}{0.20}\right] = 10\,000(0.4822531)(2.1064815) \qquad = \underline{10\,159}$$

$\qquad PV_{IN} \hspace{6cm} = \underline{\$48\,990}$

PV_{OUT}: $36\,000$ now $\hspace{5cm} = \$36\,000$

$$10\,000\left(1.20^{-3}\right) = 10\,000(0.5787037) \qquad = 5\,787$$

$$10\,000\left(1.20^{-5}\right) = 10\,000(0.4018776) \qquad = 4\,019$$

$$\left\langle 9000\left(1.20^{-7}\right)\right\rangle = \left\langle 9000(0.2790816)\right\rangle \qquad = \underline{\langle 2\,512\rangle}$$

$\qquad PV_{OUT} \hspace{5.5cm} = \underline{\$43\,294}$

$\qquad NPV \hspace{6cm} = \$\ 5\,696$

Since NPV > 0, the new product provides the required return on investment of 20% and therefore should be distributed.

7. *End of year*

	Now	1	2	3	4	5	6	7	8	9	10
In		30	30	30	30	30	30	30	30	30	30
Out	<140>				<20>			<40>			20

PV_{IN}: $30\,000\left[\dfrac{1 - 1.12^{-10}}{0.12}\right] = 30\,000(5.650223) \qquad = \$169\,507$

PV_{OUT}: $140\,000$ now $\hspace{4cm} \$140\,000$

$$20\,000\left(1.12^{-4}\right) = 20\,000(0.6355181) \qquad 12\,710$$

$$40\,000\left(1.12^{-7}\right) = 40\,000(0.4523492) \qquad 18\,094$$

$$\left\langle 20\,000\left(1.12^{-10}\right)\right\rangle = \left\langle 20\,000(0.3219732)\right\rangle \underline{\langle 6\,439\rangle} \qquad \underline{164\,365}$$

$\qquad NPV \hspace{6cm} = \$\ 5\,142$

Since NPV > 0, the investment will return more than 12% on the investment and should therefore be made.

Exercise **16.3**

A. 1.   NPV + 2350   0   −1270
           A |_____| |B
    i   24%   *d*   26%

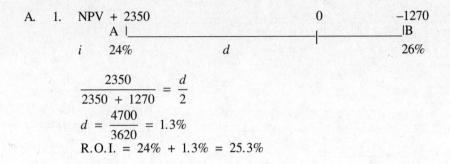

$$\frac{2350}{2350\ +\ 1270} = \frac{d}{2}$$

$$d = \frac{4700}{3620} = 1.3\%$$

R.O.I. = 24% + 1.3% = 25.3%

3.   NPV 135   0   −240
        A |_____| |B
   i   20%   *d*   22%

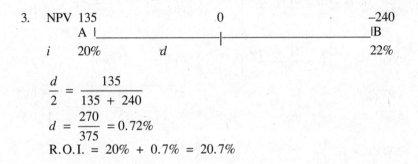

$$\frac{d}{2} = \frac{135}{135\ +\ 240}$$

$$d = \frac{270}{375} = 0.72\%$$

R.O.I. = 20% + 0.7% = 20.7%

B. 1.   *End of year*

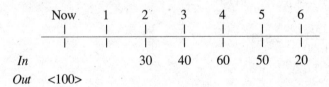

	Now	1	2	3	4	5	6
In			30	40	60	50	20
Out	<100>						

		For *i* = 14%	For *i* = 20%	For *i* = 18%
PV$_{IN}$	$30\,000(1 + i)^{-2}$	23 084	20 833	21 546
	$40\,000(1 + i)^{-3}$	26 999	23 148	24 345
	$60\,000(1 + i)^{-4}$	35 525	28 935	30 947
	$50\,000(1 + i)^{-5}$	25 968	20 094	21 855
	$20\,000(1 + i)^{-6}$	9 112	6 698	7 409
	PV$_{IN}$ =	120 688	99 708	106 102
	PV$_{OUT}$ =	100 000	100 000	100 000
	NPV	20 688	<292>	6 102

Index = $\dfrac{PV_{IN}}{PV_{OUT}}$ = 1.21 → try *i* = 20% _____↑

$$\frac{d}{2} = \frac{6102}{6102\ +\ 292}$$

$$d = \frac{12\,204}{6394} = 1.91\%$$

R.O.I. = 18% + 1.9% = 19.9%

3. *End of year*

	Now	1	2	3	4	5	6	7	8	9	10	11	12
In		2	2	2	5	5	5	5	5	3	3	3	3
Out	<15>												2

		Try $i = 16\%$	For $i = 20\%$	For $i = 18\%$
PV_{IN}	$2000\left[\dfrac{1 - (1 + i)^{-3}}{i}\right]$	4 492	4 213	4 349
	$5000(1 + i)^{-3}\left[\dfrac{1 - (1 + i)^{-5}}{i}\right]$	10 489	8 653	9 516
	$3000(1 + i)^{-8}\left[\dfrac{1 - (1 + i)^{-4}}{i}\right]$	2 561	1 806	2 147
	PV_{IN}	17 542	14 672	16 012
PV_{OUT}	15 000 now	15 000	15 000	15 000
	$\left\langle 2000(1 + i)^{-12}\right\rangle$	<337>	<224>	<274>
	PV_{OUT}	14 663	14 776	14 726
	NPV	2 879	<104>	1 286

$$\text{Index} = \frac{17\,542}{14\,663} = 1.20 \rightarrow \text{try } i = 20\%$$

$$d = \frac{2 \times 1286}{1286 + 104} = \frac{2572}{1390} = 1.85\%$$

R.O.I. $= 18\% + 1.9\% = 19.9\%$

5. *End of year*

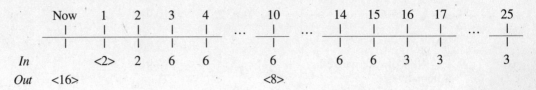

		Try $i = 16\%$	For $i = 26\%$	For $i = 22\%$	For $i = 24\%$
PV_{IN}	$\langle 2000(1 + i)^{-1} \rangle$	<1 724>	<1 587>	<1 639>	<1 613>
	$2000(1 + i)^{-2}$	1 486	1 260	1 344	1 301
	$6000(1 + i)^{-2}\left[\dfrac{1 - (1 + i)^{-13}}{i}\right]$	23 821	13 815	16 942	15 267
	$3000(1 + i)^{-15}\left[\dfrac{1 - (1 + i)^{-10}}{i}\right]$	1 565	325	596	438
	PV_{IN}	25 148	13 813	17 243	15 393
PV_{OUT}	16 000 now	16 000	16 000	16 000	16 000
	$8000(1 + i)^{-10}$	1 813	793	1 095	931
	PV_{OUT}	17 813	16 793	17 095	16 931
	NPV	7 335	<2 980>	148	<1 538>

$$\text{Index} = \frac{25\,148}{17\,813} = 1.41 \rightarrow \text{Try } 26\%$$

$$d = \frac{2 \times 148}{148 + 1538} = \frac{296}{1686} = 0.18$$

R.O.I. $= 22\% + 0.2\% = 22.2\%$

Review Exercise

1. PV of Alternative A:

$$20\,000\left(1.14^{-3}\right) = 20\,000(0.6749715) = \$13\,499$$

$$60\,000\left(1.14^{-6}\right) = 60\,000(0.4555865) = 27\,335$$

$$40\,000\left(1.14^{-10}\right) = 40\,000(0.2697438) = \underline{10\,790}$$

$$\$51\,624$$

PV of Alternative B $= 10\,000\left[\dfrac{1 - 1.14^{-10}}{0.14}\right] = 10\,000(5.2161156) = \$52\,161$

Since the PV of Alternative B > PV of Alternative A, Alternative B is preferable.

3. Alternative 1:

$$PV_{IN} = 500\left[\frac{1 - 1.03^{-36}}{0.03}\right] = 500(21.832253) = \$10\ 916$$

$$PV_{OUT} \qquad\qquad\qquad\qquad\qquad = \underline{7\ 000}$$

NPV of Alternative 1 $\qquad\qquad = \$\ 3\ 916$

Alternative 2:

$$PV_{IN} = 26\ 000\left(1.03^{-32}\right) = 26\ 000(0.388337) = \$10\ 097$$

$$PV_{OUT} \qquad\qquad\qquad\qquad\qquad = \underline{6\ 500}$$

NPV of Alternative 2 $\qquad\qquad = \$\ 3\ 597$

Since the NPV of Alternative 1 is greater than the NPV of Alternative 2, Alternative 1 is preferable.

5. *End of year*

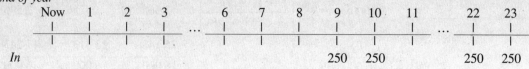

| | | | | | | | | | 250 | 250 | | 250 | 250 |

Out <75> <75> <75> <75> <75> <75>

PV_{IN} PMT = 250 000; i = 18%; n = 15; d = 8

$$PV_n(\text{defer}) = 250\ 000\left(1.18^{-8}\right)\left[\frac{1 - 1.18^{-15}}{0.18}\right]$$

$$= 250\ 000(0.2660382)(5.0915776) = \$338\ 639$$

PV_{OUT} PMT = 75 000; i = 18%; n = 8

$$PV_n(\text{due}) = 75\ 000(1.18)\left[\frac{1 - 1.18^{-8}}{0.18}\right]$$

$$= 75\ 000(1.18)(4.0775658) \qquad = \underline{360\ 865}$$

NPV $\qquad\qquad\qquad\qquad = <\$22\ 226>$

At 18%, the net present value is $-\$22\ 226$.

7.

		Try $i = 20\%$	For $i = 26\%$	For $i = 28\%$
PV_{IN}	$14\ 000\left[\dfrac{1 - (1 + i)^{-8}}{i}\right]$	53 720	45 370	43 061
PV_{OUT}	45 000	45 000	45 000	45 000
	NPV	8 720	370	<1 939>

$$\text{Index} = \frac{53\ 720}{45\ 000} = 1.19 \rightarrow \text{Try } i = 26\%$$

$$d = \frac{370 \times 2}{370 + 1939} = \frac{740}{2309} = 0.32\%$$

R.O.I. = 26% + 0.3% = 26.3%

9.

		For $i = 18\%$	For $i = 16\%$
PV_{IN}	$250\,000(1 + i)^{-8}\left[\dfrac{1 - (1 + i)^{-15}}{i}\right]$	338 639	425 164
PV_{OUT}	$75\,000(1 + i)\left[\dfrac{1 - (1 + i)^{-8}}{i}\right]$	360 865	377 892
NPV		<22 226>	47 272

$$\text{Index} = \frac{338\,639}{360\,865} = 0.938 \rightarrow \text{Try } i = 16\%$$

$$d = \frac{47\,272 \times 2}{47\,272 + 22\,226} = \frac{94\,544}{69\,498} = 1.36\%$$

R.O.I. = 16% + 1.4% = 17.4%

11. *End of year*

	Now	1	2	3	4	5	6	7	8	9	10
In		8	12	12	12	12	12	6	6	6	6
Out	<36>										9

		For $i = 20\%$	For $i = 26\%$	For $i = 24\%$
PV_{IN}	$8000(1 + i)^{-1}$	6 667	6 349	6 452
	$12\,000(1 + i)^{-1}\left[\dfrac{1 - (1 + i)^{-5}}{i}\right]$	29 906	25 096	26 568
	$6000(1 + i)^{-6}\left[\dfrac{1 - (1 + i)^{-4}}{i}\right]$	5 202	3 479	3 968
		41 775	34 924	36 988
PV_{OUT}	36 000 now	36 000	36 000	36 000
	$\left\langle 9000(1 + i)^{-10}\right\rangle$	<1 454>	<892>	<1 047>
		34 546	35 108	34 953
NPV		7 229	<184>	2 035

$$\text{Index} = \frac{41\,775}{34\,546} = 1.21 \rightarrow \text{Try } i = 26\%$$

$$d = \frac{2035 \times 2}{2035 + 184} = \frac{4070}{2219} = 1.83$$

R.O.I. = 24% + 1.8% = 25.8%

13.   Project A:

$$5800\left[\frac{1 - 1.20^{-8}}{0.20}\right] = 5800(3.8371598) = \$22\ 256$$

Project B:

$$13\ 600\left(1.20^{-1}\right) = 13\ 600(0.8333333) \quad = \$11\ 333$$

$$17\ 000\left(1.20^{-5}\right) = 17\ 000(0.4018776) \quad = \quad 6\ 832$$

$$20\ 400\left(1.20^{-8}\right) = 20\ 400(0.2325680) \quad = \quad \underline{\ \ 4\ 744}$$

$$\$22\ 909$$

Since at 20%, the PV of Project B is greater than the PV of Project A, Outway Ventures should choose Project B.

15.   *End of year*

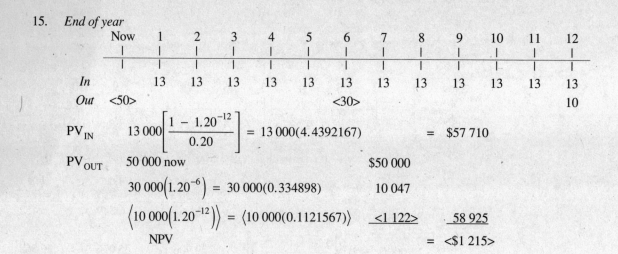

$$PV_{IN} \quad 13\ 000\left[\frac{1 - 1.20^{-12}}{0.20}\right] = 13\ 000(4.4392167) \qquad = \quad \$57\ 710$$

$$PV_{OUT} \quad 50\ 000\ now \qquad\qquad\qquad\qquad\quad \$50\ 000$$

$$30\ 000\left(1.20^{-6}\right) = 30\ 000(0.334898) \qquad 10\ 047$$

$$\left\langle 10\ 000\left(1.20^{-12}\right)\right\rangle = \left\langle 10\ 000(0.1121567)\right\rangle \quad \underline{<1\ 122>} \qquad \underline{\ \ 58\ 925}$$

$$NPV \qquad\qquad\qquad\qquad\qquad\qquad = \ <\$1\ 215>$$

17. *End of year*

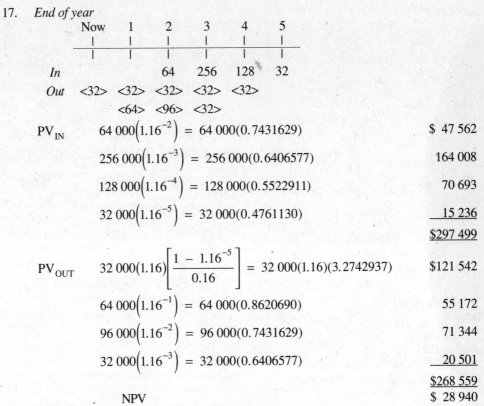

$$PV_{IN} \quad 64\,000\left(1.16^{-2}\right) = 64\,000(0.7431629) \qquad \$\ 47\,562$$

$$256\,000\left(1.16^{-3}\right) = 256\,000(0.6406577) \qquad 164\,008$$

$$128\,000\left(1.16^{-4}\right) = 128\,000(0.5522911) \qquad 70\,693$$

$$32\,000\left(1.16^{-5}\right) = 32\,000(0.4761130) \qquad \underline{15\,236}$$

$$\$297\,499$$

$$PV_{OUT} \quad 32\,000(1.16)\left[\frac{1 - 1.16^{-5}}{0.16}\right] = 32\,000(1.16)(3.2742937) \qquad \$121\,542$$

$$64\,000\left(1.16^{-1}\right) = 64\,000(0.8620690) \qquad 55\,172$$

$$96\,000\left(1.16^{-2}\right) = 96\,000(0.7431629) \qquad 71\,344$$

$$32\,000\left(1.16^{-3}\right) = 32\,000(0.6406577) \qquad \underline{20\,501}$$

$$\$268\,559$$

NPV $\qquad \$\ 28\,940$

Since at 16%, the NPV is positive, the return on investment will be greater than 16% and the product should be marketed.

Self-Test

1. Present value of Alternative A:

$$2500\left[\frac{1 - 1.15^{-12}}{0.15}\right] = 2500(5.420619) \ = \ \$13\,552$$

Present value of Alternative B:

$$10\,000(1.15)^{-4} = 10\,000(0.5717532) \ = \ \$\ 5\,718$$

$$10\,000(1.15)^{-8} = 10\,000(0.3269018) \ = \ \ \ 3\,269$$

$$10\,000(1.15)^{-12} = 10\,000(0.1869072) \ = \ \underline{\ \ 1\,869}$$

$$\$10\,856$$

At 15%, PV(A) > PV(B).
Preferred alternative is A.

3. *End of year*

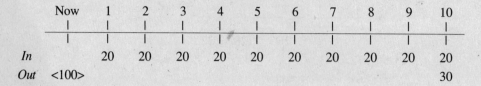

		For $i = 12\%$	For $i = 18\%$	For $i = 16\%$
PV_{IN}	$20\,000\left[\dfrac{1 - (1 + i)^{-10}}{i}\right]$	113 004	89 882	96 665
PV_{OUT}	100 000 now	100 000	100 000	100 000
	$\left\langle 30\,000(1 + i)^{-10}\right\rangle$	<9 659>	<5 732>	<6 801>
		90 341	94 268	93 199
	NPV	22 663	<4 386>	3 466

$\text{Index} = \dfrac{113\,004}{90\,341} = 1.25 \rightarrow \text{Try } i = 18\%$

$d = \dfrac{3466}{3466 + 4386} \times 2 = \dfrac{6932}{7852} = 0.8849\%$

R.O.I. = 16% + 0.9% = 16.9%

5. Proposal A:

$PV_{IN} \quad 20\,000\left[\dfrac{1 - 1.20^{-10}}{0.20}\right] \qquad\qquad \$83\,849$

$PV_{OUT} \quad$ Immediate outlay $\qquad\qquad \$60\,000$

After 3 years: $40\,000(1.20)^{-3}$ $\qquad \underline{23\,148} \qquad \underline{83\,148}$

$\qquad\qquad$ NPV(A) $\qquad\qquad\qquad\qquad \$ \quad 701$

Proposal B:

$PV_{IN} \quad 40\,000\left[\dfrac{1 - 1.20^{-7}}{0.20}\right](1.20)^{-3} \qquad \$83\,440$

$PV_{OUT} \quad 29\,000(1.20)\left[\dfrac{1 - 1.20^{-4}}{0.20}\right] \qquad \$90\,088$

Less $50\,000(1.20)^{-10}$ $\qquad\qquad \underline{<8\,075>} \qquad \underline{82\,013}$

$\qquad\qquad$ NPV(B) $\qquad\qquad\qquad\qquad \$ \; 1\,427$

Since NPV(B) > NPV(A), Proposal B is preferred at 20%.